Rupinder Singh, Ranvijay Kumar, Vinay Kumar, J. Paulo Davim (Eds.)
3D Printing

De Gruyter Series in Advanced Mechanical Engineering

Series Editor
J. Paulo Davim

Volume 10

3D Printing

Polymer, Metal and Gel Based Additive Manufacturing

Edited by
Rupinder Singh, Ranvijay Kumar, Vinay Kumar
and J. Paulo Davim

DE GRUYTER

Editors
Dr. Rupinder Singh
Teachers Training and Research
National Institute of Technical
Sukhna Path, Sector 26
Chandigarh 160019
India
rupindersingh@nitttrchd.ac.in

Dr. Ranvijay Kumar
University Center for Research and Development
Chandigarh University
NH-05, Ludhiana, Chandigarh State Hwy
Mohali 140413
India
ranvijayk12@gmail.com

Dr. Vinay Kumar
University Center for Research and Development
Chandigarh University
NH-05, Ludhiana, Chandigarh State Hwy
Mohali 140413
India
vinaykumar6192@gmail.com

Prof. Dr. J. Paulo Davim
Department of Mechanical Engineering
University of Aveiro
Campus Santiago
3810-193 Aveiro
Portugal
pdavim@ua.pt

ISBN 978-3-11-108061-1
e-ISBN (PDF) 978-3-11-108163-2
e-ISBN (EPUB) 978-3-11-108206-6
ISSN 2367-3796

Library of Congress Control Number: 2024930002

Bibliographic information published by the Deutsche Nationalbibliothek
The Deutsche Nationalbibliothek lists this publication in the Deutsche Nationalbibliografie; detailed bibliographic data are available on the internet at http://dnb.dnb.de.

Cover image: Sky_blue/iStock/thinkstock
Typesetting: Integra Software Services Pvt. Ltd.
Printing and binding: CPI books GmbH, Leck

MIX
Papier | Fördert gute Waldnutzung
FSC www.fsc.org FSC® C083411

www.degruyter.com

Contents

Abhishek Barwar*, Prateek Kala, Rupinder Singh

1 3D-printed smart functional prototypes as sensors and actuators for robotic applications

Abstract: Soft robotics, inspired by nature, is a trending research area that is based on material science and the available mechanism (pneumatic, thermal, magnetic, electrostatic, etc.) to activate them. In the recent past, polylactic acid (PLA) has shown promising material characteristics, and researchers explored the 4D printing of PLA-based functional prototypes that can be used for high-strength biomedical applications for bovines. But hitherto, little has been explored about the sensing and actuation capabilities of PLA for implantable robotic applications in bovine. In this study, PLA extracted from biomedical waste has been used to prepare the functional prototypes (implants in bovine). Further, it has been primarily recycled by varying number of shredding cycles, and finally, feedstock filaments of the same have been prepared on the single screw extruder. For analyzing the sensing and actuation capabilities of the recycled material, the prototypes were fabricated on the fused deposition modeling setup using the following process parameters: raster angle, 45°; infill pattern, grid; printing speed, 40 mm/s; and infill density, 50%. Initially, the pressure-based sensor was designed in the high-frequency structure simulator simulation software and then tested on the vector network analyzer setup to observe the dielectric properties, that is, dielectric constant (ϵ_r) and loss tangent ($\tan \delta$) under the effect of pressure. Further, a hydrothermal actuator for robotic application has been programmed by mechanical force, and the shape recovery of the deformed prototype has been observed under the effect of hydrothermal stimuli. The change in dielectric properties (ϵ_r and $\tan \delta$) of the recycled material has been observed through the insertion loss (S_{21}) versus resonant frequency (f_r), which may be used for online health monitoring of patients through a Bluetooth device by observing the electromagnetic signal in the industrial, scientific, and medical (ISM) band.

Acknowledgments: This study was supported by NITTTR, Chandigarh (India), BITS Pilani, and GADVASU, Ludhiana (India). Financial support by DST (file no. SR/FST/COLLEGE/2020/997) is also acknowledged for this project.

***Corresponding author: Abhishek Barwar**, Department of Mechanical Engineering, BITS Pilani, Pilani, Jhunjhunu, Rajasthan, India, e-mail: barwarabhishek@gmail.com
Prateek Kala, Department of Mechanical Engineering, BITS Pilani, Pilani, Jhunjhunu, Rajasthan, India
Rupinder Singh, Department of Mechanical Engineering, National Institute of Technical Teachers Training and Research, Chandigarh, India

https://doi.org/10.1515/9783111081632-001

Graphical abstract:

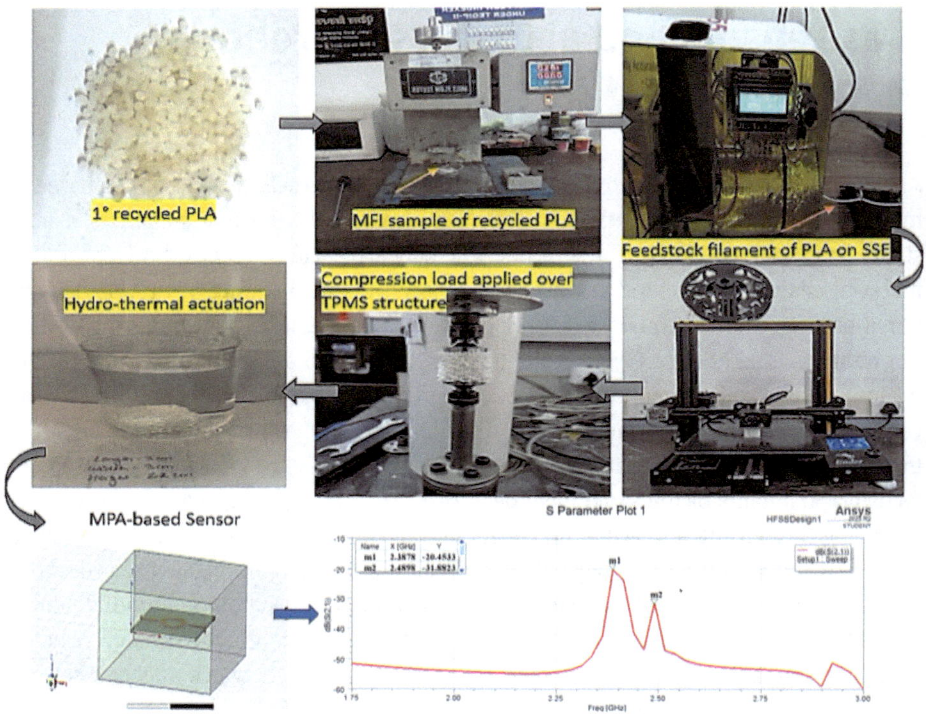

Keywords: Soft robotics, actuation, dielectric constant, loss tangent, ISM band

1.1 Introduction

Livestock farming plays an important role in the rural areas of any nation and provides a source of income to the families living in villages. Buffaloes being an integral part of the bovine family contribute to the society in terms of milk production, hides and draft power, meat production, and so on (Coroian et al. 2013). Diaphragmatic hernia (DH) is a severe disorder commonly observed in milk buffaloes (due to the absence of pressure receptors) (Athar et al. 2010). DH may be caused due to improper feeding habits, ingestion of foreign bodies, traumatic situations, violent falls, and so on (Mostafa et al. 2015). DH may lead to patient's death in some cases if the patient is not provided treatment on time (Talekar et al. 2018). Rumenotomy is a surgical procedure to clean the rumen and is generally performed before doing the herniorrhaphy (a surgical process to fix the damaged diaphragm) following a post-xiphoid approach (Pooniya et al. 2022). Reoccurrence of DH is a common problem in buffaloes' postoperative DH surgery which may be because of the ingestion of foreign bodies, and during the advanced stage of pregnancy, it may weaken the diaphragm progressively (Kumar et al. 2012).

Soft robotics is a trending field for implantable biomedical devices, actuators, sensors, and so on (Keneth et al. 2021). These are mainly based on material science and can be driven by different actuating mechanisms such as pneumatic (Morin et al.2012), thermal (Zhang et al. 2011), and magnetic (Jin et al. 2018; Schmauch et al. 2017). Implantable robotics mainly consists of three subunits: sensor, actuator, and the powering mechanism to run the system (Keneth et al. 2021). Additive manufacturing (AM) is a class of manufacturing processes that are used to produce complex components (Singh et al. 2022) while dealing with a wide variety of materials in the form of stacking layer over layer (Singh et al. 2020). In anatomical research, 3D printing plays a crucial role such that it can provide a patient-specific product and replicate the complex anatomical features in a precise manner (Kurenov et al. 2015). In the recent past, 3D printing has expanded its application in the development of soft robots and implantable robotic devices (Gul et al. 2018). Fused filament fabrication/fused deposition modeling (FDM) is one of the AM techniques that was initially used for prototyping, and nowadays used extensively for biomedical applications such as the fabrication of implants and scaffolds (Jayanth et al. 2018). It generally deals with thermoplastics such as acrylonitrile butadiene styrene, polylactic acid (PLA), and polycaprolactone (Prakash et al. 2021). PLA is a bioderived and biodegradable polymer and is suitable for biomedical applications due to its biocompatible nature (Ranjan et al. 2020). The combination of AM with intelligent materials, also known as 4D printing, leads to the production of such components that can change their shape or volume upon exposure to an external stimuli (Griffin 2019). PLA often shows some inherent shape memory properties (SMPs) when exposed to temperature usually beyond glass transition temperature (T_g) and may be used for soft robotics applications (Leist et al. 2017). The semicrystalline structure and hydrophobic nature of PLA make it an excellent choice (due to its biodegradable nature) for fabricating sensors for biomedical applications (Silvers et al. 2012). In addition to the above, the SMPs of PLA also make it eligible for the development of sensing devices for conformal applications (Kumar et al. 2020).

Microstrip patch antennas (MPAs) have been used extensively in biomedical applications such as biotelemetry and detection of cancer tumors (Ridoy et al. 2021). The MPA-based sensor has three subunits that unite to make a single assembly (i.e., a patch, substrate, and ground plane). In the recent past, a smart mesh in the sensor form (based on the principle of MPA) has been proposed to detect the recurrence of DH (Indian patent no.: 202311036753). Apart from conformability, an Internet of things-based solution may be helpful for the online health monitoring of such patients. During the past decade, researchers explored the 4D printing of PLA-based functional prototypes that may be used for high-strength biomedical applications. But hitherto, little has been explored about the sensing and actuation capabilities of PLA for implantable robotic applications. For further identification of the research gap, a detailed bibliometric analysis (BA) has been performed in this work using innovative bibliometric tools such as VOSviewer, RStudio, and Biblioshiny.

1.2 Bibliometric analysis

In this study, a detailed BA has been performed to analyze the trending research topics and the amount of research required in a particular area (in the form of graphs/network diagrams). Certain keywords, that is, "robotics," "biomedical application," and "sensors and actuators" have been searched on the Web of Science (WOS) platform, and based upon that, a total of 97 documents were retrieved in the form of (plain text file) from the period of 2004–2023 (Tab. 1.1) which represents that limited research has been reported in the field. The annual growth rate of the research in this area has been observed as 14.45%, which is relatively low, whereas the average citation per document was 44.84. Other information like author, document, and publication has been mentioned in Tab. 1.1.

Tab. 1.1: Main information about the data extracted from the WOS platform.

Description	Results
Main information about data	
Time span	2004:2023
Sources (journals, books, etc.)	56
Documents	97
Annual growth rate %	14.45
Document average age	3.64
Average citations per doc	44.84
References	8,746
Document contents	
Keywords plus (ID)	373
Author's keywords (DE)	331
Authors	
Authors	502
Authors of single-authored docs	2
Authors' collaboration	
Single-authored docs	2
Co-authors per doc	5.47
International co-authorships %	27.84
Document types	
Article	65
Article; early access	2
Article; proceedings paper	2
Review	28

1.2.1 Co-occurrence analysis

The co-occurrence analysis of different terms (extracted from the WOS database) has been performed in the freeware "VOSviewer." This analysis provides information on the interconnections between various terms by analyzing the co-occurrence of two terms in more than one article. To prepare the network diagram, the minimum number of appearances of a term in an article has been selected to be 3, and it has been observed that 52 out of 671 terms match the threshold criteria. Further, the strength of the co-occurrence links along with all other keywords was calculated in the software for each of the 52 keywords, and based on the co-occurrence link strength, the network diagram has been prepared. Figure 1.1(a) represents the network visualization diagram based on the keywords "robotics," "sensors and actuators," and "biomedical applications." The terms were divided into five different color clusters randomly. Similarly, an overlay visualization graph of the analysis has been highlighted in Fig. 1.1(b). Trending terms over the last 5 years have been represented by different colors for each of the years, for example, terms such as "robot-sensing systems," "shape memory polymers," and "4D printing" were among the most frequently reported terms in the year 2022.

The co-occurrence map upon highlighting the term "3D printing" is shown in Fig. 1.1(c), and it has been observed that 3D printing has connections with emerging fields like soft robotics and strain sensors but does not have a connection with the term "robot sensing system" that highlights gap in the research. Similarly, highlighting the term "sensor" (Fig. 1.1(d)), it has been noticed that it does not have any co-appearance along with the terms SMP, medical robots, and systems and represents the need for research in the field. Figure 1.1(e) represents the limited use of SMP for implantable robotic applications and represents an area of research to work on.

1.2.2 Three-field plot

The three-field plot, also known as the Sankey diagram, provides information about the interrelation between three unidentical fields. Each of the fields is represented by a rectangular block expressed by different color codes, and the larger the block size, the larger the number of interconnections, a field that will have the other two. Figure 1.2 represents the interconnections between "source," "author country," and "keywords." It was observed that countries like China and the USA have interconnections with each source and keyword, whereas countries like India and Korea only have connections with limited keywords like 4D printing and soft robotics and hence highlights the need for efforts for trending fields like robot sensing and actuators. Figure 1.2 also highlights that journal like *Sensors and Actuators A: Physical* has proven prominent (for such research areas) by the fact that it has been connected with almost every country.

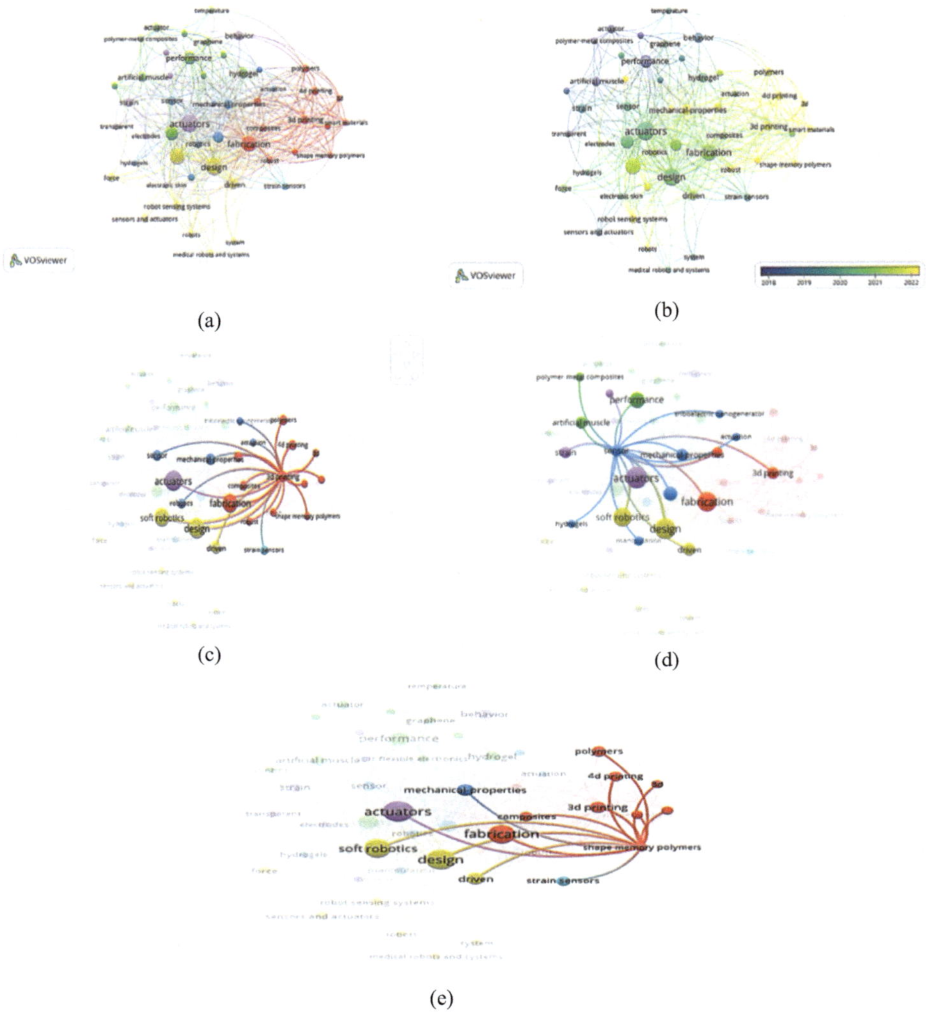

Fig. 1.1: Co-occurrence analysis based on keywords "3D printing," "sensors," and "SMPs"; (a) network visualization diagram, (b) overlay visualization, (c) co-occurrence of terms along with "3D printing," (d) interconnections between "sensor" and other terms; and (e) "SMP" and its interconnection with other terms.

1.2.3 Trend topics

In this study, the topics explored in the fields of 3D printing, robotic applications, and sensors or actuators have been identified for the past two decades. Figure 1.3 shows that terms like artificial muscle, biomedical applications, and fabrication have shown tremendous growth recently and hence highlight the need for time. Figure 1.4 represents the word cloud map showcasing the most explored words in the past and cur-

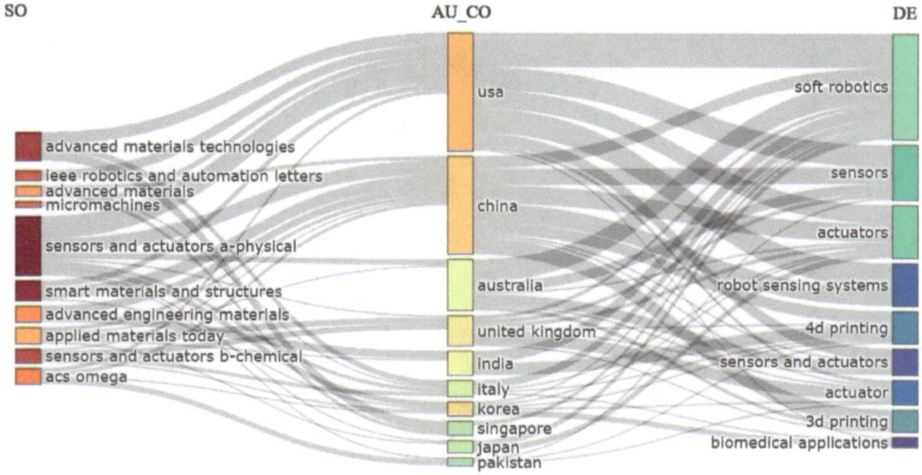

Fig. 1.2: Three-field plots based on the field: source, country, and keyword.

rent and has been shown by different sizes as per their frequency of occurrence (i.e., the larger the size, the maximum frequency it will have).

1.3 Materials and methodology

PLA was selected as a material to fabricate the sensor and actuator in this study. PLA's biocompatible and biodegradable nature may be responsible for its wide use in biomedical applications. Some of the properties like easy to process, availability, and printability make it a suitable material for 3D printing. In this study, PLA having a density of 1.25 g/cm^3 has been used to fabricate the smart functional prototypes via a 3D printing technique named FDM. Since PLA possesses some SMPs under the effect of external stimuli, it is suitable for the manufacturing of smart actuators and soft robotic applications in the biomedical field. The methodology (Fig. 1.5) involves certain steps: (i) selection of material; (ii) 3D printing of part of functional prototypes; (iii) fabrication of smart sensor; (iv) programming of a smart actuator; and (v) actuation of the smart prototype under a controlled stimulus.

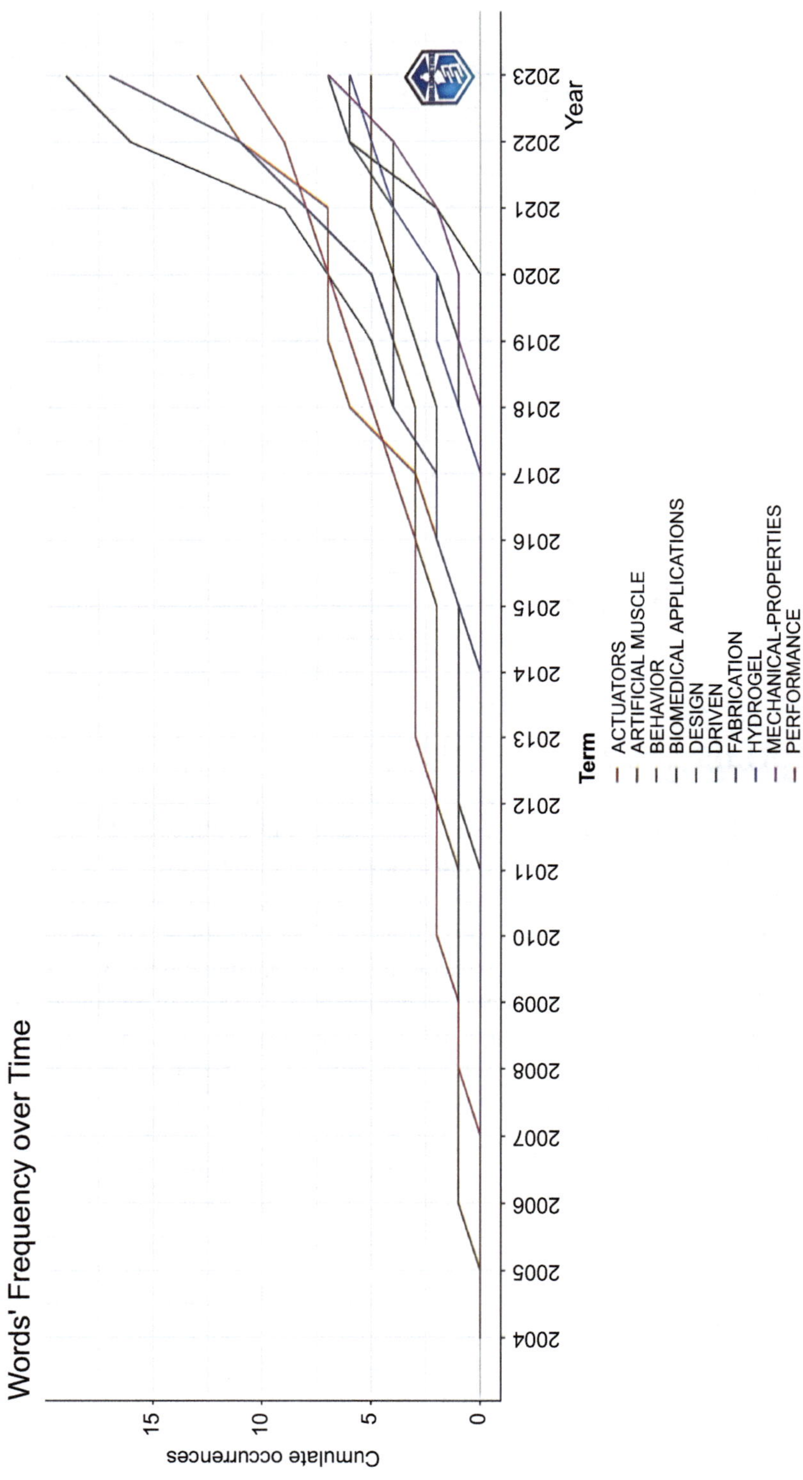

Fig. 1.3: Trending keywords over the past two decades as per their cumulative occurrences.

Fig. 1.4: Word cloud map based on the cumulative frequency score.

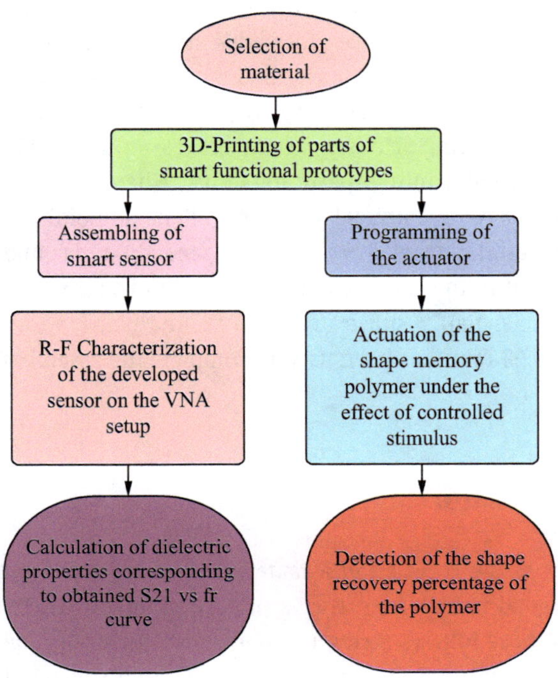

Fig. 1.5: Detailed methodology followed in this study.

1.4 Experimentation

1.4.1 Rheological investigation

To analyze the flow properties of the PLA, its melt flow index (MFI) has been performed at the following parameters: temperature, 185 °C; applied load, 2.160 kg as per ASTM D1238. To ensure the MFI with minimum error percentage, the experiment has been repeated thrice, and the average value of the same has been considered for the 3D printing of functional prototypes. The average value of MFI of the recycled PLA has been identified as 11.967 g/10 min and has shown repeatability. As per the literature, the MFI value in the higher range was found to be suitable for the fabrication of thin sections and hence desirable for the sensing application.

1.4.2 Preparation of feedstock filaments and functional prototypes

The recycled PLA extracted from biomedical waste in the form of granules has been processed through a single screw extruder as per the MFI results. The parameters selected for filament preparation are as follows: temperature, 180 °C; screw speed, 7 rpm; and filament diameter, 1.75 mm. The prepared feedstock strands were then used for the FDM printing of substrate for the smart sensor and a prototype for mimicking the actuation capabilities of the material. The substrate has dimensions of 78.40 mm × 80 mm fabricated on the following printing parameters: raster angle, 45°; infill pattern: octet; infill density, 50%; and printing speed: 100 mm/s. Whereas to perform the actuation behavior, a triply periodic minimal structure (TPMS, lidinoid) has been printed at parametric settings: infill density, 50%; printing speed, 40 mm/s; and infill pattern: lines. As the sensor used in this study comprised three units such as a metallic patch (to radiate/transfer the signal), a substrate, and a ground plane, a thin adhesive film of copper (thickness: 0.08 mm) has been used to complete the assembly (Fig. 1.6).

1.4.3 Design of sensor and actuator

To identify the sensing capabilities of the recycled PLA material, a sensor based on the principle of MPA has been designed in high-frequency structure simulator (HFSS) software (Fig. 1.7). The sensor used here follows a ring resonator test to identify the dielectric properties of the substrate material. It comprises three subunits: ground plane, substrate, and a patch (a ring and two feed lines). To calculate the dimensions of these three units, some of the established mathematical expressions (1.1)–(1.6) (Singh et al. 2022) have been used in this study:

(a) (b)

Fig. 1.6: 3D-printed functional prototypes: (a) MPA-based sensor and (b) TPMS lidinoid.

$$2\pi R = n\lambda_g \tag{1.1}$$

where R is defined as the mean ring radius and λ_g is the guided wavelength:

$$\lambda_g = \frac{c}{\sqrt{\varepsilon_{\text{eff}}}} \times \frac{1}{f_r} \tag{1.2}$$

Here c is the velocity of light, ϵ_{eff} is the effective dielectric constant of the material, and f_r is the frequency at which the sensor shows its peak value:

$$\varepsilon_{\text{eff}} = \left[\frac{\varepsilon_r + 1}{2}\right] + \left[\left(\frac{\varepsilon_r - 1}{2}\right)\left(1 + 12\frac{h}{w}\right)^{-0.5}\right] \tag{1.3}$$

Here h and w are the substrate height and width of the ring and microstrip lines, respectively, ε_r is the dielectric constant:

$$L = \frac{\lambda_g}{4} \tag{1.4}$$

Here L is the length (quarter of guided wavelength) of feed lines

$$R_i = R - \frac{w}{2} \tag{1.5}$$

$$R_o = R + \frac{w}{2} \tag{1.6}$$

Here R_i and R_o represent the inner and outer ring radius values, respectively, whereas w is the width of microstrip lines

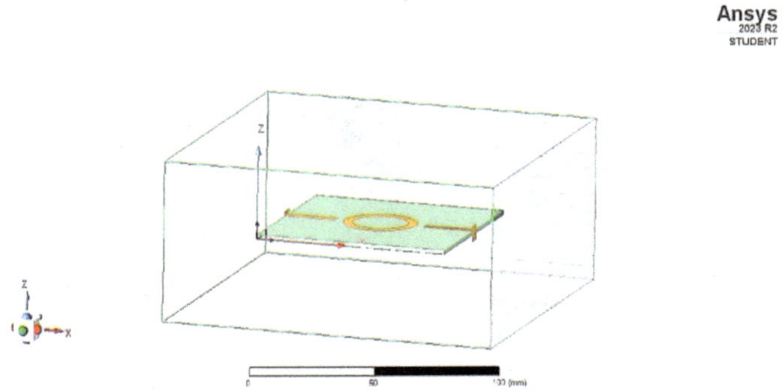

Fig. 1.7: Sensor designed in the HFSS software as per the calculations.

1.4.4 Programming of actuator

To analyze the behavior of the actuator, it has been programmed mechanically in this study to ascertain the shape memory effect. The fabricated TPMS sample has been tested on an UTM, and a compression load has been applied over it at a deformation rate of 1 mm/min at room temperature. The initial dimension of a square block was 30 mm × 30 mm, and a deformation of 8 mm was provided via a compression test (Fig. 1.8).

(a) (b)

Fig. 1.8: Programming of the actuator: (a) compressive load applied over the structure and (b) deformed shape of the sample.

1.5 Results and discussion

1.5.1 R-F characteristics of the sensor

The R-F characteristics of the fabricated sensor have been measured on the vector network analyzer (VNA) setup within the range of 1–4 GHz (Tab. 1.2).

Tab. 1.2: Observations from the simulation and measurement results of the sensor.

Simulated result	Measured result
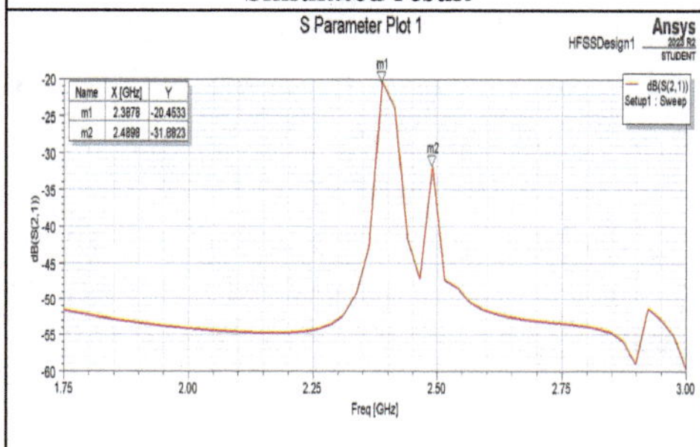	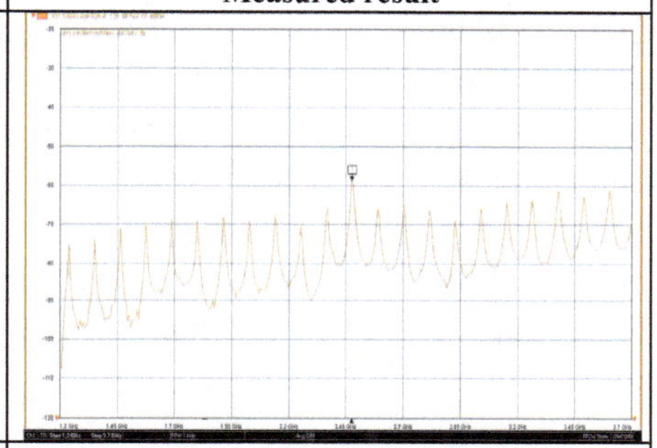
$S_{21} = -31.8823$ dB, $f_r = 2.4898$ GHz	$S_{21} = -58.7003$, $f_r = 2.4768$ GHz

Before testing the sensor, the calibration of the VNA has been performed based on three standards (open, short, and through). The sensor consists of two ports and has been tested for S_{21} versus f_r, and at peak value, the dielectric properties have been calculated using equations (1.7) and (1.8):

$$\varepsilon_r = \frac{(2 \times \varepsilon_{eff} + M - 1)}{M + 1} \tag{1.7}$$

Here M represents the fringing effects on the edges of microstrip surfaces and also a function of the height of the substrate

$$\tan \delta = \frac{\alpha_d \lambda_o}{27.3} \left(\frac{(\varepsilon_r - 1)\sqrt{\varepsilon_{eff}}}{\varepsilon_r (\varepsilon_{eff} - 1)} \right) \tag{1.8}$$

Here α_d and λ_o represent the attenuation factor of the substrate material and wavelength of the resonant signal.

Going through these calculations, the ε_r and $\tan \delta$ have been calculated as 2.6454 and 0.0044, respectively. The fabricated sensor gives a signal in the industrial, scientific, and medical (ISM) band (at 2.47 GHz) and is hence suitable for biomedical applications. The dielectric properties of the substrate material can change upon deformation in shape and hence the sensor can resonate at some other frequency. The increase or decrease in load over the substrate material can be measured with the change in dielectric properties. To ensure repeatability in the results, its actuation behavior has been analyzed.

1.5.2 Actuation of the TPMS sample

To further analyze the actuation capabilities of the material, a hydrothermal actuation was found to be suitable for activating the SMPs of PLA. The actuation of the sample has been performed at 65 °C above the glass transition temperature (T_g) of PLA in a temperature-controlled water tank for 8 h. Shape recovery of the SMP sample has been observed using a measuring scale is shown in Fig. 1.9. It was noticed that the deformed structure regains its length up to 28 mm (i.e., 73.33%). Since temperature and water act as an actuating mechanism, more length recovery may be observed with long exposure conditions. Since PLA is a biocompatible material and also shows promising SMP results, it may be used for high-strength biomedical applications, where repetitive loading conditions that exist can be used as an extensive unit for medical robots (like in DH surgery).

(a) (b)

Fig. 1.9: Shape recovery observations: (a) deformed length of the sample, that is, 22 mm, and (b) recovered length after stimulus effect, that is, 28 mm.

1.6 Conclusions

This study highlights the sensing and actuation capabilities of primarily recycled PLA (extracted from biomedical waste) for robotic applications. Flow characteristics (MFI-11.967 g/10 min) of the material suggest that it is good for the fabrication of thin sections. The results of the actuation test suggest that the material has excellent SMPs as it showed a tremendous 73.33% recovery in shape under the effect of hydrothermal stimulus and hence can be used as an extensive unit of robots for high-strength implantable applications. For analyzing the sensing behavior, an MPA-based sensor has been designed in HFSS as well as fabricated via FDM, and the results, that is, both simulated and VNA, have very little deviation and hence capable of measuring very small deformations as well and give the signal in the form of change in dielectric properties of the material upon deformation. The simulated values of dielectric properties are $S_{21} = -31.8823$ dB and $f_r = 2.4898$ GHz, and the values measured on VNA are $S_{21} = -58.7003$ and $f_r = 2.4768$ GHz. The SMP properties of PLA make it a useful material option for fabricating such sensors for implantable robotics and biomedical applications.

For further research, piezoresponsive materials may be explored, and load-based stimuli can be effective for them as the limitation of PLA is its T_g which is higher than the body temperature and hence less suitable for actuation purposes.

References

Athar H, Mohindroo J, Singh K, Kumar A, Raghunath M. Comparison of radiography and ultrasonography for diagnosis of diaphragmatic hernia in bovines. Veterinary Medicine International. 2010 Oct;2010.

Coroian A, Erler S, Matea CT, Mireșan V, Răducu C, Bele C, Coroian CO. Seasonal changes of buffalo colostrum: Physicochemical parameters, fatty acids, and cholesterol variation. Chemistry Central Journal. 2013 Dec;7:1–9.

Griffin J. The State of Creativity: The Future of 3D Printing, 4D Printing and Augmented Reality. Edward Elgar Publishing, 2019.

Gul JZ, Sajid M, Rehman MM, Siddiqui GU, Shah I, Kim KH, Lee JW, Choi KH. 3D printing for soft robotics–a review. Science and Technology of Advanced Materials. 2018 Dec 31;19(1):243–62.

Indian Patent- A post-surgery detecting device for diaphragmatic hernia in bovines, Application No.: 202311036753, filed on Dated: 27/05/2023.

Jayanth N, Senthil P, Prakash C. Effect of chemical treatment on tensile strength and surface roughness of 3D-printed ABS using the FDM process. Virtual and Physical Prototyping. 2018 Jul 3;13(3):155–63.

Jin B, Song H, Jiang R, Song J, Zhao Q, Xie T. Programming a crystalline shape memory polymer network with thermo-and photo-reversible bonds toward a single-component soft robot. Science Advances. 2018 Jan 26;4(1):eaao3865.

Kumar A, Singh Saini N, Mohindroo J, Sangwan V, Kant Mahajan S, Raghunath M, Singh N. Long-term outcomes of survivors of diaphragmatic herniorrhaphy in crossbred cows and buffaloes. Indian Journal of Animal Sciences. 2012 Sep 1;82(9):971.

Kumar R, Singh R, Singh M, Kumar P. On ZnO nanoparticle reinforced PVDF composite materials for 3D printing of biomedical sensors. Journal of Manufacturing Processes. 2020 Dec 1;60:268–82.

Kurenov SN, Ionita C, Sammons D, Demmy TL. Three-dimensional printing to facilitate anatomic study, device development, simulation, and planning in thoracic surgery. The Journal of Thoracic and Cardiovascular Surgery. 2015 Apr 1;149(4):973–79.

Leist SK, Gao D, Chiou R, Zhou J. Investigating the shape memory properties of 4D printed polylactic acid (PLA) and the concept of 4D printing onto nylon fabrics for the creation of smart textiles. Virtual and Physical Prototyping. 2017 Oct 2;12(4):290–300.

Morin SA, Shepherd RF, Kwok SW, Stokes AA, Nemiroski A, Whitesides GM. Camouflage and display for soft machines. Science. 2012 Aug 17;337(6096):828–32.
Mostafa MB, Abu-Seida AM, Abdelaal AM, Al-Abbadi OS, Abbas SF. Ultrasonographic features of the reticulum in normal and hardware diseased buffaloes. Research Opinions in Animal and Veterinary Sciences. 2015;5(4):165–71.
Pooniya R, Palsania SK, Saini R, Kumar A, Lal M, Kumar H. Surgico-therapeutic management of diaphragmatic hernia in buffalo: A review of 8 cases.
Prakash C, Senthil P, Sathies T. Fused deposition modeling fabricated PLA dielectric substrate for microstrip patch antenna. Materials Today: Proceedings. 2021 Jan 1;39:533–37.
Ranjan N, Singh R, Ahuja IP. Development of PLA-HAp-CS-based biocompatible functional prototype: A case study. Journal of Thermoplastic Composite Materials. 2020 Mar;33(3):305–23.
Ridoy PM, Elme KM, Saha P, Hoque MJ, Tulka TK, Rahman MA. Rectangular microstrip patch antenna for biomedical application using ISM band. In 2021 International Conference on Intelligent Technologies (CONIT). IEEE, 2021 Jun 25, 1–6.
Sachyani Keneth E, Kamyshny A, Totaro M, Beccai L, Magdassi S. 3D printing materials for soft robotics. Advanced Materials. 2021 May;33(19):2003387.
Schmauch MM, Mishra SR, Evans BA, Velev OD, Tracy JB. Chained iron microparticles for directionally controlled actuation of soft robots. ACS Applied Materials & Interfaces. 2017 Apr 5;9(13):11895–901.
Silvers AL, Chang CC, Parrish B, Emrick T. Strategies in aliphatic polyester synthesis for biomaterial and drug delivery applications. In: Degradable Polymers and Materials: Principles and Practice (2nd Edition). American Chemical Society, 2012, 237–54.
Singh R, Das A, Anand A. On 3D printing of customized multi-root dental implants for the strategic tooth of canine by direct metal laser sintering. Rapid Prototyping Journal. 2023 Mar 2;29(3):447–59.
Singh R, Kumar S, Singh AP, Wei Y. On comparison of recycled LDPE and LDPE–bakelite composite based 3D printed patch antenna. Proceedings of the Institution of Mechanical Engineers, Part L: Journal of Materials: Design and Applications. 2022 Apr;236(4):842–56.
Singh T, Kumar S, Sehgal S. 3D printing of engineering materials: A state of the art review. Materials Today: Proceedings. 2020 Jan 1;28:1927–31.
Talekar SH, Chaudhary MP, Gamiti KS, Katara KP, Manish S. Surgical management of diaphragmatic hernia in four Jaffarabadi buffaloes and two Gir cows. International Journal of Environmental Science and Technology. 2018;7(2):684–88.
Zhang X, Pint CL, Lee MH, Schubert BE, Jamshidi A, Takei K, Ko H, Gillies A, Bardhan R, Urban JJ, Wu M. Optically-and thermally-responsive programmable materials based on carbon nanotube-hydrogel polymer composites. Nano Letters. 2011 Aug 10;11(8):3239–44.

Minhaz Husain*, Rupinder Singh, Bahadur Singh Pabla, Seema Ramniwas, Ranvijay Kumar

2 Biomimetic-based 3D-printed smart implants

Abstract: In the past decade, several studies have been reported on polyvinylidene fluoride (PVDF) and its composites for energy storage device applications. Also, some studies have reported the use of PVDF composite as a useful material for the fabrication of implants. But hitherto the detailed procedure for in-house development of PVDF–hydroxyapatite (HAp)–chitosan (CS)-based biomimetic 3D-printed smart implant at the lab level has not been reported. This study highlighted the mechanical properties (tensile, flexural, and compressive) and biocompatibility features based on in vitro testing for the in-house development of PVDF composite. The result of the study suggests that the PVDF composite-based biomimetic 3D-printed smart implants may be used in orthopedic applications for online health monitoring based on the concept of microstrip patch antenna for bone growth. The observation of the biocompatibility test of the PVDF composite shows that the best combination of HAp and CS in the PVDF matrix was noticed in sample 3, that is, the 90% PVDF–8% HAp–2% CS (for the egg-to-adult survival of *Drosophila melanogaster*). The composition (90% PVDF–8% HAp–2% CS) was used to fabricate the biomimetic-based 3D-printed smart implant. The result suggests that from a multifactor optimization viewpoint, 235 °C nozzle temperature, 40 mm/s printing speed, 60° raster angle (suggested by the regression model), and 90° raster angle (suggested by a general linear model) are the best setting of fused filament fabrication.

Acknowledgments: This study was supported by NITTTR, Chandigarh (India), and Panjab University, Chandigarh, for laboratory support.

***Corresponding author: Minhaz Husain**, Department of Mechanical Engineering, National Institute of Technical Teachers Training and Research, Chandigarh, India, e-mail: minhazme2012@gmail.com
Rupinder Singh, Bahadur Singh Pabla, Department of Mechanical Engineering, National Institute of Technical Teachers Training and Research, Chandigarh, India
Seema Ramniwas, Ranvijay Kumar, University Centre for Research and Development, Chandigarh University, Mohali, Punjab, India

https://doi.org/10.1515/9783111081632-002

Graphical abstract:

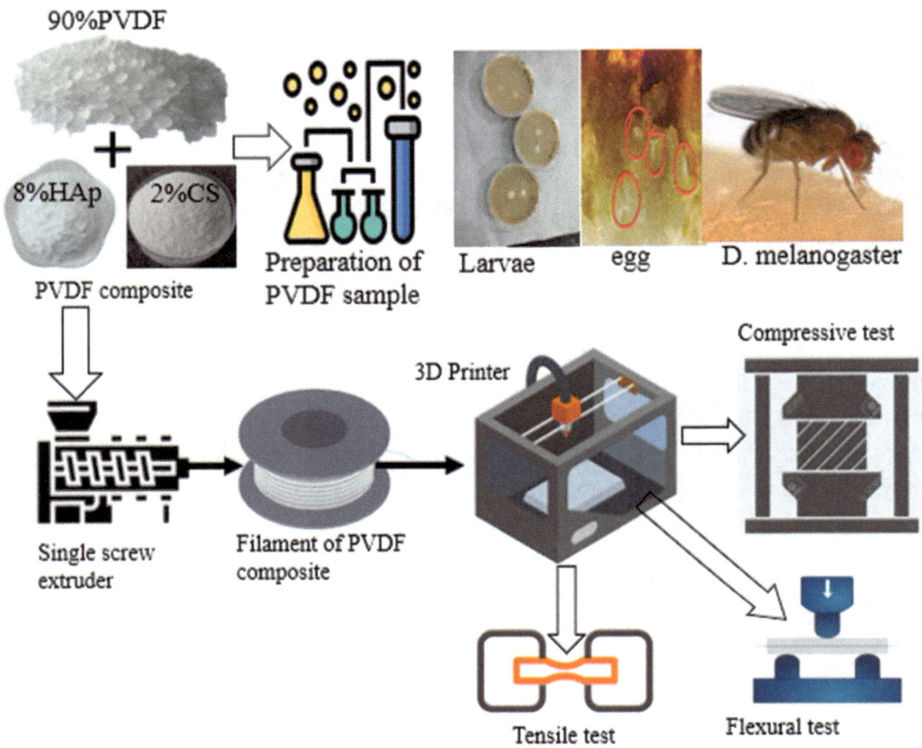

Keywords: Biomimetic materials, 3D printing, biocompatibility, smart implants

2.1 Introduction

A medical device or prosthesis that mimics the shape and functionality of organic tissues or organs in the human body is known as a biomimetic implant (Singh et al. 2023a–c). Biomimetic implants are designed to increase the overall performance while reducing the chance of rejection and improving body compatibility. They aim to mimic the biological characteristics and the activity of live tissues by drawing inspiration from nature (Chen and Liu 2016). Usually, biomimetic implants are created using the same design principles as natural biological systems. This can entail imitating the design, dynamics, or material composition of particular tissues or organs. These implants may be constructed by a biocompatible polyvinylidene fluoride (PVDF) composite, so the body would not reject them or have an unpleasant immunological reaction to them (Husain et al. 2023a–d). With biomimetic implants, the functioning of the tissue or an organ (they are replacing) is improved or restored. For instance, biomimetic heart valves mimic the way natural heart valves work to improve blood flow and lower the risk of problems (Shao et al. 2021). These implants can aid in lowering inflammation and the chance of infection, which are major worries with traditional prostheses or implants since they closely resemble natural tissues. The integration of biomimetic implants with the surrounding tissues is frequently facilitated. This may result in a longer implant life span as well as more natural mobility and function. Biomimetic implants are used in many different medical specialties. Examples include biomimetic bone implants that replicate the characteristics and structure of natural bone for orthopedic purposes (Singh et al. 2023d) and biomimetic retinal implants that mimic the function of the retina to restore vision in situations of retinal degeneration (Weiland and Humayun 2005). The development of novel materials, production methods, and designs by researchers to enhance the functionality and durability of these devices has resulted in a constant evolution of the area of biomimetic dental implants (Singh et al. 2023b). It may be noted that biomimetic implants have numerous benefits, yet they may also have drawbacks and difficulties, such as the difficulty of accurately simulating specific biological processes and the possibility of long-term wear and tear. To progress in the area of biomimetic implants and enhance patient outcomes, researchers are still working to solve these problems.

Additive manufacturing, often known as 3D printing, enables the accurate and adaptable creation of complex items, including implants for use in medicine (Husain et al. 2023a). The capacity to customize the design to the patient's unique demands is one of the main benefits of 3D-printed implants. This personalization guarantees a better fit and function for the person, which may result in better results. Complex implant designs may be made with 3D printing, which might not be possible with conventional production techniques (Husain et al. 2023c). This is especially advantageous for implants with intricate internal systems or those that must match the patient's particular anatomy. Implant designs may be rapidly prototyped and iterated via 3D printing, which can hasten the development process and cut expenses. Some implants created through 3D printing are made of bioresorbable materials, which eventually degrade (Husain et al. 2023e) and are absorbed by the body. For temporary implants or those that provide medications or therapies, this may be useful. Biocompatible metals (such as titanium or cobalt–chromium alloys), polymers, ceramics, and composite materials may be utilized to 3D-print implants (Kumar et al. 2023a). The particular application and required qualities will choose the material to use. Orthopedic operations like hip and knee replacements frequently involve 3D-printed implants. By precisely matching the patient's bone structure to the implants, surgeons can improve results and minimize problems (Kumar et al. 2023b). Dental implants, crowns, bridges, and orthodontic appliances may all be made specifically for a patient using 3D printing (Singh et al. 2023a). Tissue engineering uses 3D printing to make scaffolds and constructs for regenerating or repairing damaged tissues and organs in addition to hard implants. Similar to conventional implants, metallic 3D-printed medical implants (Pradhan et al. 2022) must go through regulatory approval procedures in many nations. A crucial part of getting these gadgets on the market is ensuring their safety, effectiveness, and quality. It has been reported that the development of novel materials, and printing processes, for 3D-printed implants is constantly changing. This field of innovation has the potential to completely change the healthcare sector. These include the need to resolve regulatory barriers and guarantee the mechanical and biocompatibility of the printed materials (Morrison et al. 2015). The capabilities of 3D-printed implants and their potential to enhance patient care, however, are still being expanded as a result of continuous research and development in the area.

A medical device or prosthesis that was constructed utilizing 3D printing and other sophisticated technologies, such as sensors, electronics, and data processing capabilities, is known as a 3D-printed smart implant (Husain et al. 2023b). This implant is made to track and communicate with the patient's body, deliver real-time data, and react to shifting circumstances. Numerous sensors that can track variables like temperature, pressure, pH levels, and biological indicators are frequently included in smart implants (Andreu et al. 2015). Both the patient and the healthcare practitioners can benefit from the data that these sensors can supply. The ability to wirelessly communicate with other devices, such as smartphones, tablets, or medical monitoring systems, is a feature found in many smart implants (Kadhim et al. 2020). Real-time feedback and remote monitoring are made possible by this implant. Some intelligent implants are made to deliver drugs or healing substances right to the place that needs them. These implants can release medications in response to certain physiological cues or signals that are sent from a distance. Biocompatible materials are generally used in the construction of smart implants to reduce the possibility of tissue rejection or undesirable responses (Ledet et al. 2018). In different medical specialties, including neurology for brain stimulation, cardiology for monitoring heart issues, and orthopedics for joint monitoring and rehabilitation, these novel implants may be employed. As researchers investigate new technologies and applications, the area of 3D-printed smart implants is constantly changing. The capabilities of these implants are also improving with the incorporation of artificial intelligence and machine learning. A promising development in healthcare is the use of 3D-printed smart implants, which might lead to more individualized and efficient medical treatments, real-time monitoring, and better patient outcomes (Javaid et al. 2022). However, ethical, statutory, and safety considerations must be given serious thought throughout the creation and application of such implants.

2.2 Background and problem formulation

Bibliometric analysis was performed using the Web of Science database for the past 20 years. A total of 81 relevant research papers were discovered on the Web of Science database when the keyword "biomimetic-based 3D-printed implant" was explicitly explored. To further enhance the study, these articles were examined using the open VOSviewer application with a minimal occurrence requirement of 5. Out of a total of 3,267 terms, 86 terms that met the set threshold criteria were shortlisted. The top 60% of pertinent terms (52) were selected. As stated in Tab. 2.1, each of the 52 terms had a relevance score calculated to identify which are most important. These 52 terms were chosen for the network diagram depicted in Fig. 2.1. For the keyword "3D printing," a research gap analysis was also carried out; the outcomes are depicted in Fig. 2.2. The highlighted region in Fig. 2.2 represents the study on 3D-printed implants. However, there were no nodes for a 3D-printed smart implant in the reviewed data.

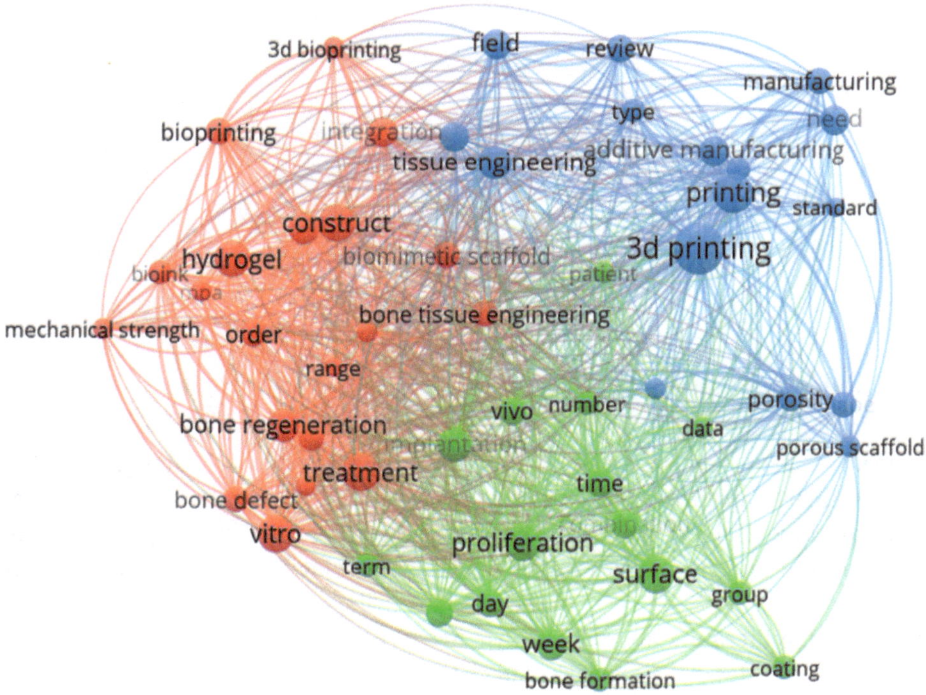

Fig. 2.1: Network diagram for the keyword "biomimetic-based 3D-printed implant."

A bibliometric study of the last 20 years on the topic of "biomimetic-based 3D-printed smart implant" was also explored using the Web of Science database. There were three relevant result outcomes. Open-source VOSviewer software was used to further improve the study, taking a minimum occurrence criterion of 2. A total of 134 terms from the recovered terms satisfied the requirements. A relevance score was determined for each of the four words, as shown in Tab. 2.2. The most important terms were chosen using this score, with the top 60% of relevant terms being chosen by default. As a result, four words were selected to construct the network diagram shown in Fig. 2.3. Additionally, a research gap assessment for the term "scaffolds" was per-

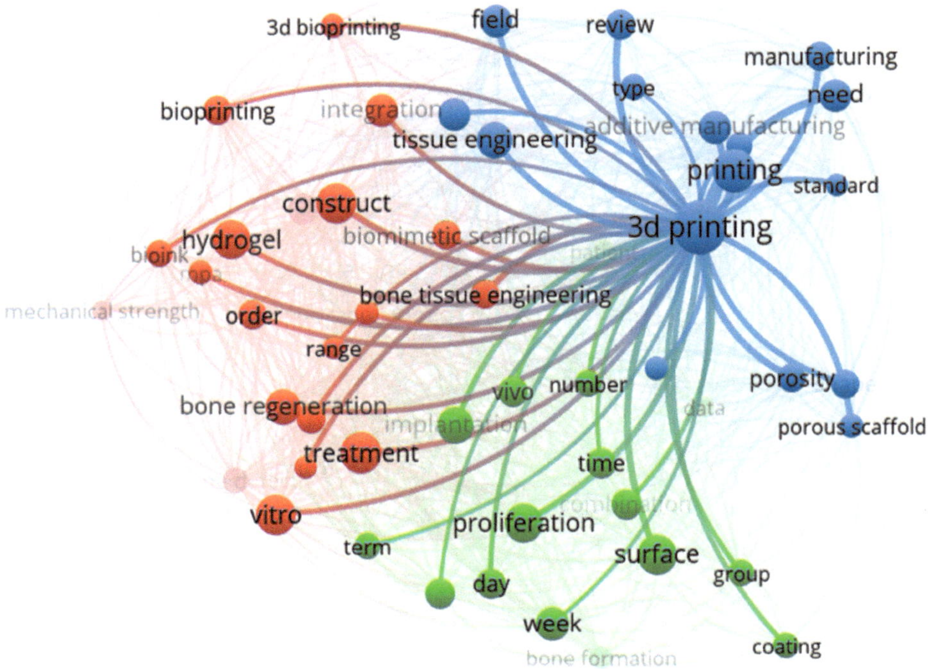

Fig. 2.2: Research gap for the keyword "3D-printed implant."

formed; the results are shown in Fig. 2.4. In the investigation of biomimetic-based 3D-printed smart implants, it is noteworthy that no nodes were discovered. These findings offer insightful information on the state of the art of research and point to possible directions for further study in the domain of biomimetic-based 3D-printed smart implants.

From the literature and bibliometric network analysis, it has been ascertained that in the past decade, number of studies have been reported on PVDF and its composites for energy storage device applications. Also, some studies have reported the use of PVDF–hydroxyapatite (HAp)–chitosan (CS) composite as a useful material for the fabrication of implants. But hitherto the detailed process for in-house development of PVDF composite-based biomimetic 3D-printed smart implant at the lab level is not reported. This study is the extension of previous work on PVDF composite sensors for biomedical applications (Husain et al. 2023c). Overall this study highlights the biocompatibility test of PVDF with varying weight percentages of reinforcement of HAp and CS. After that, the mechanical properties (tensile, flexural, and compressive) for the in-house development of PVDF composite-based biomimetic 3D-printed smart implant at a lab scale have been established.

Tab. 2.1: Occurrence score for the keyword biomimetic-based 3D-printed implant.

Id	Term	Occurrences	Relevance score
1	3D bioprinting	6	1.5209
2	3D printing	26	0.8974
3	Additive manufacturing	10	1.8845
4	Aspect	7	1.305
5	Bioink	7	1.2296
6	Biomimetic scaffold	8	0.2881
7	Bioprinting	8	1.4606
8	Bone defect	8	0.8889
9	Bone formation	7	1.556
10	Bone regeneration	12	0.5862
11	Bone tissue engineering	8	0.3795
12	Clinical application	6	0.6945
13	Coating	7	1.9649
14	Combination	9	0.6548
15	Composite scaffold	8	0.4171
16	Construct	15	0.5226
17	Data	5	0.6803
18	Day	9	0.5642
19	Extracellular matrix	11	0.878
20	Field	10	1.1431
21	Group	7	1.5294
22	Hydrogel	14	0.8269
23	Implantation	13	0.4274
24	Integration	10	0.5172
25	Manufacturing	8	3.0374
26	Mechanical strength	5	1.1556
27	MPA	6	0.6556
28	Mu m	5	0.6066
29	Need	10	3.1826
30	Number	7	0.269
31	Order	8	0.3886
32	Osteogenesis	9	0.6771
33	Patient	6	0.5555
34	Polycaprolactone	6	0.7347
35	Porosity	8	1.0634
36	Porous scaffold	6	1.6886
37	Printing	18	1.2007
38	Proliferation	14	0.277
39	Range	5	0.6196
40	Review	9	1.973
41	Shape	8	1.36
42	Standard	5	1.7653
43	Surface	15	0.7331
44	Term	7	0.666
45	Time	9	0.4068
46	Tissue engineering	13	0.5682

Tab. 2.1 (continued)

Id	Term	Occurrences	Relevance score
47	Tissue regeneration	10	1.2865
48	Treatment	16	0.6578
49	Type	7	0.9318
50	In vitro	15	0.8057
51	In vivo	9	0.5101
52	Week	11	1.4066

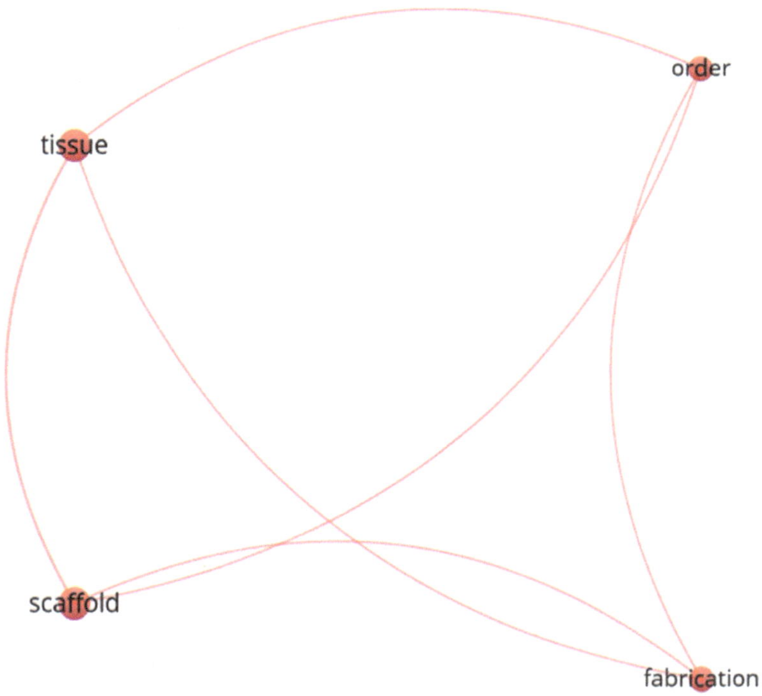

Fig. 2.3: Network diagram for the biomimetic-based 3D-printed smart implant.

2.3 Materials and methods

A PVDF extrusion grade was purchased from Solvay, Gujarat (India). The pellet has a density of 1.75 g/cm^3, a melting point between 170 and 180 °C, and a glass transition temperature (T_g) of −40 °C. The Marine Hydrocolloids at Kochi, Kerala, India, provided CS (degree of deacetylation 90%, pH 7.0–9.0) as well as HAp (colorless and brittle). Based on published studies, the composition and ratio of PVDF matrix were chosen. Figure 2.5 outlines the study's approach in detail.

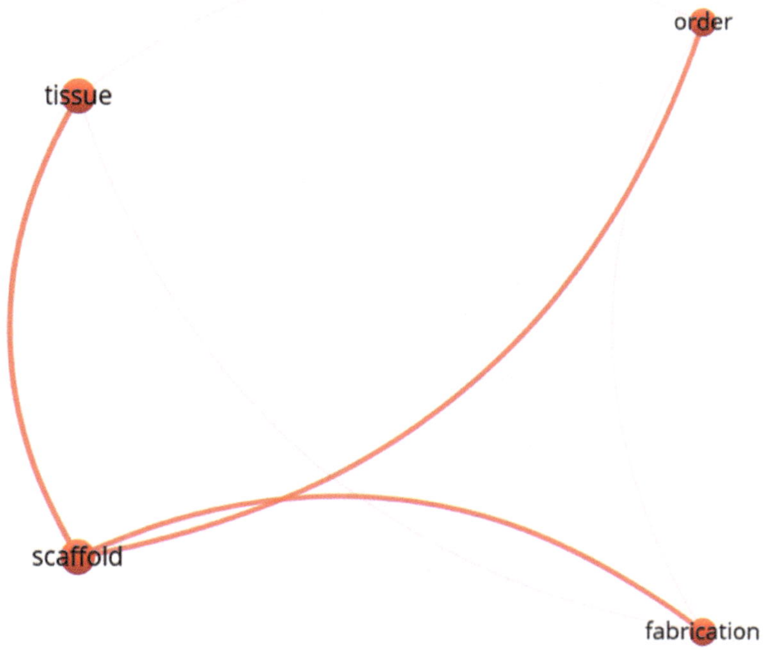

Fig. 2.4: Research gap for the keyword "scaffolds."

Tab. 2.2: Occurrence score for the keyword "biomimetic-based 3D-printed smart implant."

Id	Term	Occurrences	Relevance score
1	Fabrication	2	0.627
2	Order	2	0.627
3	Scaffold	3	1.373
4	Tissue	3	1.373

2.4 Experimentation

In this section, initially the biocompatibility test of PVDF composite with varying weight percentages of reinforcement (HAp–CS) with PVDF matrix was performed. After that, the smart implant was fabricated and tested for mechanical properties.

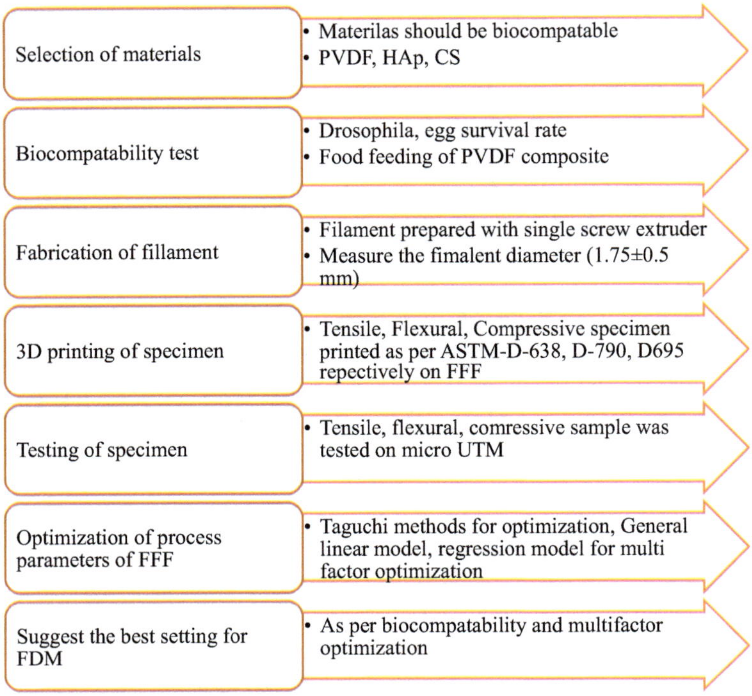

Fig. 2.5: Adopted methodology.

2.4.1 Biocompatibility test

The specimen was prepared with varying weight percentages of PVDF matrix with reinforcement of HAp and CS. The four weight percentages were selected as per the literature reported. The four compositions of matrix and its reinforcements are: S1, 100% PVDF; S2, 90% PVDF–10% HAp; S3, 90% PVDF–8% HAp–2% CS; and S4, 90% PVDF–9% HAp–1% CS.

2.4.2 Methods for preparation of PVDF surface for the biocompatibility experiment

Drosophila were exposed to the surface of PVDF for 24 h (Nasser et al. 2016). *Drosophila melanogaster* was allowed to lay eggs in a Petri dish containing control media (maize powder 72 g, sugar 64 g, agar 15 g, yeast powder 40 g, propionic acid 6 mL, and sodium benzoate 3 g) of *Drosophila* with different PVDF surfaces. Flies not exposed to PVDF were considered controls in this study. For each concentration of PVDF, five replicates were prepared and incubated in the biochemical oxygen demand at 25 °C.

A design of experiment as per Taguchi L9 for three factors with three levels shown in Tab. 2.3 was selected for the optimization of input process parameters of fused filament fabrication (FFF).

Tab. 2.3: Factor information.

Factor	Levels	Values
Nt (°C)	3	225, 230, 235
Ps (mm/s)	3	40, 50, 60
Ra (°)	3	0, 45, 90

Note: Nt, nozzle temperature; Ps, printing speed; Ra, raster angle.

2.5 Results and discussion

2.5.1 Toxicity tests

For ecotoxicological tests, eggs of *Drosophila melanogaster* were exposed to the surface of PVDF, and all tests were performed five times to check the larval mortality. Mean egg-to-adult development time along with survivorship was calculated as follows:

$$\text{Egg-to-adult survival} = \frac{\text{Number of adults surviving} \times 10}{\text{Total number of eggs}} \ldots \tag{2.1}$$

The PVDF surface exposure to *D. melanogaster* does not show mortality at the phenotypic level in preadult and adult stages. The egg-to-adult survival of *D. melanogaster* on the PVDF surface was similar to the egg-to-adult development of control lines, which contain normal food (Figs. 2.6 and 2.7). PVDF showed negligible toxic effects on *D. melanogaster*. The egg-to-adult survival for control food and 100% PVDF was observed to be the same in sample 2 (90% PVDF–10% HAp) and was observed near the same generation of egg-to-adult survival. Sample 3 (90% PVDF–8% HAp–2% CS) showed a higher number of egg-to-adult survival, but in sample 4 (90% PVDF–9% HAp–1% CS), the percentage of egg-to-adult survival was found to be less than in sample 3. The observation of the biocompatibility test of the PVDF composite shows that the best combination of HAp and CS in the PVDF matrix was found in sample 3, that is, 90% PVDF–8% HAp–2% CS for the egg-to-adult survival of *D. melanogaster*.

2.5.2 Result of mechanical properties

Analysis of variance (ANOVA), which is a statistical method for examining the causes of variation in processing and identifying if there are statistically significant differen-

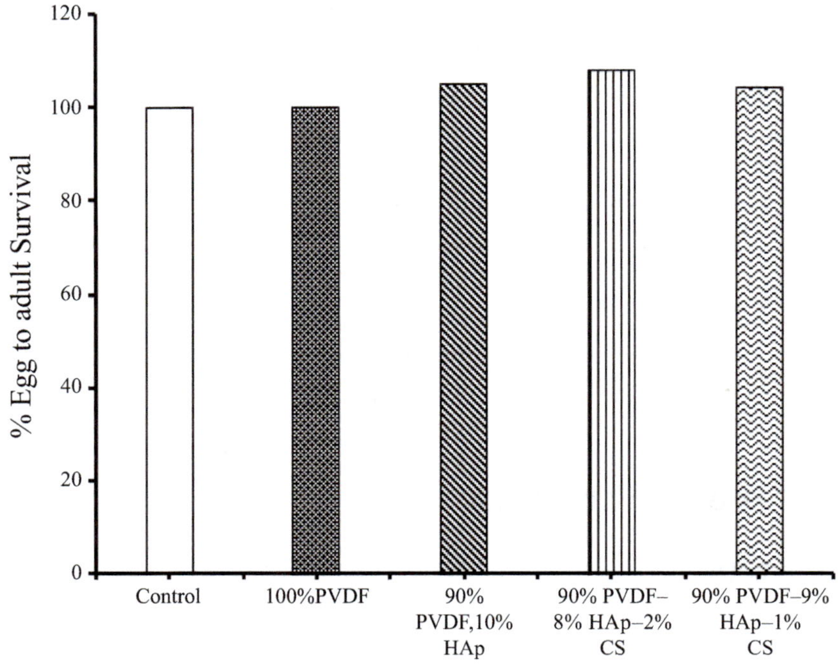

Fig. 2.6: Egg-to-adult survival.

Fig. 2.7: Petri dishes with different weight percentages of PVDF composite specimen.

ces across selected variables, was used in this study. An ANOVA table (Tab. 2.5) may be used to evaluate the effects of input process parameters on the output parameters (peak load for tensile, flexural, and compressive properties given in Tab. 2.4) of the printed specimens.

Tab. 2.4: Input and output responses for the optimization of FFF.

S. no.	Nt (°C)	Ps (mm/s)	Ra (°)	Peak load for tensile (N)	Peak load for flexural (N)	Peak load for compressive (N)	S/N for tensile	S/N for flexural	S/N for compressive
1	225	40	0	274.219	293.94	904.2	48.762	49.365	59.125
2	225	50	45	197.084	228.84	1295.5	45.893	47.191	62.249
3	225	60	90	287.048	256.38	1019.4	49.159	48.178	60.167
4	230	40	45	377.672	385.52	716.4	51.542	51.721	57.103
5	230	50	90	169.118	395.42	563.4	44.564	51.941	55.016
6	230	60	0	319.2	261.04	1019.4	50.081	48.334	60.167
7	235	40	90	394.872	407.01	1,321	51.929	52.192	62.418
8	235	50	0	393.202	308.34	1,270	51.892	49.781	62.076
9	235	60	45	393.202	263.22	500.6	51.892	48.406	53.990

Note: Five repetitions were performed, and the average value was noted, S/N, signal-to-noise ratio.

Tab. 2.5: ANOVA for S/N ratios.

Mechanical characteristics	ANOVA for S/N ratios						
	Source	DF	Seq. SS	Adj. SS	Adj. MS	*F*	*P*
Tensile	Nt (°C)	2	26.443	26.443	13.222	2.85	0.260
	Ps (mm/s)	2	19.562	19.562	9.781	2.11	0.322
	Ra (°)	2	4.593	4.593	2.296	0.49	0.669
	Residual error	2	9.290	9.290	4.645		
	Total	8	59.888				
Flexural	Nt (°C)	2	9.6928	9.6928	4.8464	11.13	0.082
	Ps (mm/s)	2	11.6560	11.6560	5.8280	13.39	0.069
	Ra (°)	2	5.3661	5.3661	2.6831	6.16	0.140
	Residual error	2	0.8706	0.8706	0.4353		
	Total	8	27.5855				
Compressive	Nt (°C)	2	14.823	14.823	7.411	0.31	0.765
	Ps (mm/s)	2	4.927	4.927	2.464	0.10	0.907
	Ra (°)	2	10.751	10.751	5.376	0.22	0.818
	Residual error	2	48.316	48.316	24.158		
	Total	8	78.817				

Note: DF, degree of freedom; Seq. SS, sequential square of sum; Adj. SS, adjusted square of sum; Adj. MS, adjusted mean square; *F*, Fisher's value; *P*, probability.

The term "source of variation" in the context of FFF optimization refers to the elements or variables that might affect the caliber of 3D printing. These sources may include print parameters (nozzle temperature (Nt), printing speed (Ps), and raster angle (Ra)), material characteristics (such as filament type and moisture content), or even environmental variables (such as relative humidity and temperature). The amount of data variability that can be linked to a certain component or source of variation is measured in terms of degrees of freedom. The ratio of the mean square within groups to the mean square between groups is known as the *F*-value. It reveals if the difference between group variance is statistically significant. A high *F*-value in the context of FFF optimization denotes that at least one factor, such as Nt, Ps, and Ra, significantly affects mechanical properties.

Tab. 2.6: Parameters of the general linear model.

Response	Goal	Lower	Target	Upper	Weight	Importance
Peak load for compressive (N)	Maximum	500.600	1,321.00		0.6	1
Peak load for flexural (N)	Maximum	228.840	407.01		0.3	2
Peak load for tensile (N)	Maximum	169.118	394.87		0.1	3

Tab. 2.7: Solution for the general linear model.

Solution	Nt (°C)	Ps (mm/s)	Ra (°)	Peak load for compressive (N) fit	Peak load for flexural (N) fit	Peak load for tensile (N) fit	Composite desirability
1	235	40	90	1,065.69	419.126	402.889	0.963405

The *p*-value represents the likelihood that the result will be seen if there is no actual difference between the groups (i.e., if all changes are the result of random chance). The *p*-value (usually less than 0.05) indicates that the input process parameters under study were significant. In this case, the *p*-value was greater than 0.05 so that it goes for the multifactor optimization of process parameters of FFF.

2.5.3 General linear model

For peak load in tensile, flexural, and compressive properties of PVDF–HAp–CS, a general linear model (Tab. 2.6) was applied. The solution of the general linear model is given in Tab. 2.7, and the composite desirability of the study is shown in Fig. 2.8. Figure 2.8 depicts that the composite desirability of the process parameters was more than 95%. The weightage of mechanical properties is selected based on the force acting on the bone during daily life.

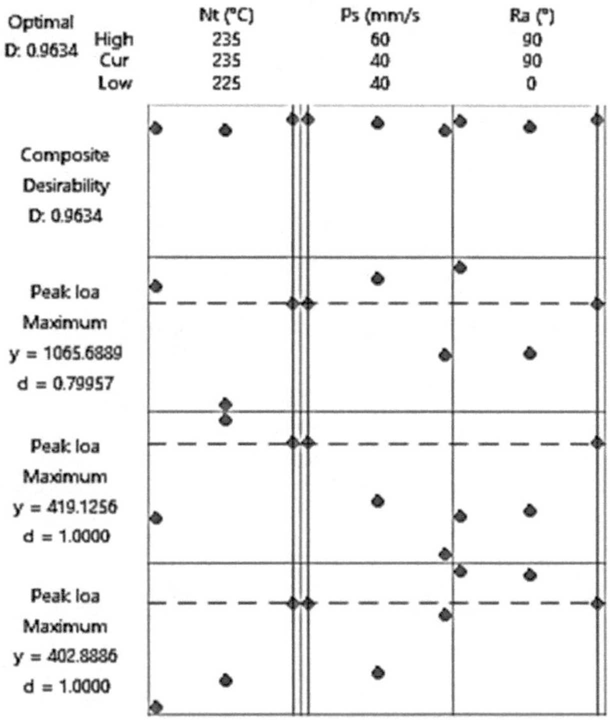

Fig. 2.8: Plot of composite desirability of the general linear model.

2.5.4 Regression analysis

Regression analysis was also performed for the peak of tensile, flexural, and compressive properties of the PVDF composite-based implant prototype. The regression equation of various mechanical properties is shown in the following sections.

2.5.4.1 Regression equations for tensile properties

$$\text{Peak load for tensile (N)} = -2,869 + 14.10\,\text{Nt}\,(^\circ\text{C}) - 0.79\,\text{Ps}\,(\text{mm/s}) - 0.502\,\text{Ra}\,(^\circ)$$

$$\text{Peak strain for tensile (\%)} = -71.0 + 0.391\,\text{Nt}\,(^\circ\text{C}) - 0.137\,\text{Ps}\,(\text{mm/s}) - 0.0373\,\text{Ra}\,(^\circ)$$

$$E\ \text{for tensile (MPa)} = -0.50 + 0.0081\,\text{Nt}\,(^\circ\text{C}) + 0.0176\,\text{Ps}\,(\text{mm/s}) - 0.00031\,\text{Ra}\,(^\circ)$$

$$\text{MoT for tensile (MPa)} = -2.304 + 0.01197\,\text{Nt}\,(^\circ\text{C}) - 0.00187\,\text{Ps}\,(\text{mm/s}) + 0.000289\,\text{Ra}\,(^\circ)$$

2.5.4.2 Regression equations for flexural properties

$$\text{Peak load for flexural (N)} = -995 + 6.65\,\text{Nt}\,(^\circ\text{C}) - 5.10\,\text{Ps}\,(\text{mm/s}) + 0.724\,\text{Ra}\,(^\circ)$$

$$\text{Peak stress for flexural (MPa)} = -3 + 0.28\,\text{Nt}\,(^\circ\text{C}) - 0.631\,\text{Ps}\,(\text{mm/s}) + 0.081\,\text{Ra}\,(^\circ)$$

$$\text{Stress at break for flexural (MPa)} = -99.2 + 0.406\,\text{Nt}\,(^\circ\text{C}) + 0.2129\,\text{Ps}\,(\text{mm/s}) + 0.0115\,\text{Ra}\,(^\circ)$$

$$\text{Extension at load for flexural (mm)} = 0.424 - 0.00157\,\text{Nt}\,(^\circ\text{C}) - 0.000867\,\text{Ps}\,(\text{mm/s}) - 0.000259\,\text{Ra}\,(^\circ)$$

2.5.4.3 Regression equations for compressive properties

$$\text{Peak load for compressive (N)} = 2,318 - 4.3\,\text{Nt}\,(^\circ\text{C}) - 6.7\,\text{Ps}\,(\text{mm/s}) - 1.07\,\text{Ra}\,(^\circ)$$

$$\text{Peak elongation for compression (mm)} = 23.09 - 0.0840\,\text{Nt}\,(^\circ\text{C}) - 0.0280\,\text{Ps}\,(\text{mm/s}) - 0.00544\,\text{Ra}\,(^\circ)$$

$$\text{Peak load for compressive (N)} = 2,318 - 4.3\,\text{Nt}\,(^\circ\text{C}) - 6.7\,\text{Ps}\,(\text{mm/s}) - 1.07\,\text{Ra}\,(^\circ)$$

$$\text{Elongation at peak (\%) for compression} = 55.3 - 0.2000\,\text{Nt}\,(^\circ\text{C}) - 0.0667\,\text{Ps}\,(\text{mm/s}) - 0.01111\,\text{Ra}\,(^\circ)$$

The parameters of regression analysis are given in Tab. 2.8, the solution of regression analysis is given in Tab. 2.9, and the composite desirability of the regression analysis is shown in Fig. 2.9. Figure 2.9 depicts that the composite desirability of the process parameters was more than 95%. The weightage of mechanical properties is selected based on the force acting on the bone during daily life.

Tab. 2.8: Parameters for regression analysis.

Response	Goal	Lower	Target	Upper	Weight	Importance
Peak load for compressive (N)	Maximum	500.600	1,321.00		0.6	1
Peak load for flexural (N)	Maximum	228.840	407.01		0.3	2
Peak load for tensile (N)	Maximum	169.118	394.87		0.1	3

Tab. 2.9: Solution for regression analysis.

Solution	Nt (°C)	Ps (mm/s)	Ra (°)	Peak load for compressive (N) fit	Peak load for flexural (N) fit	Peak load for tensile (N) fit	Composite desirability
1	235	40	60.9091	985.363	406.804	382.119	0.945883

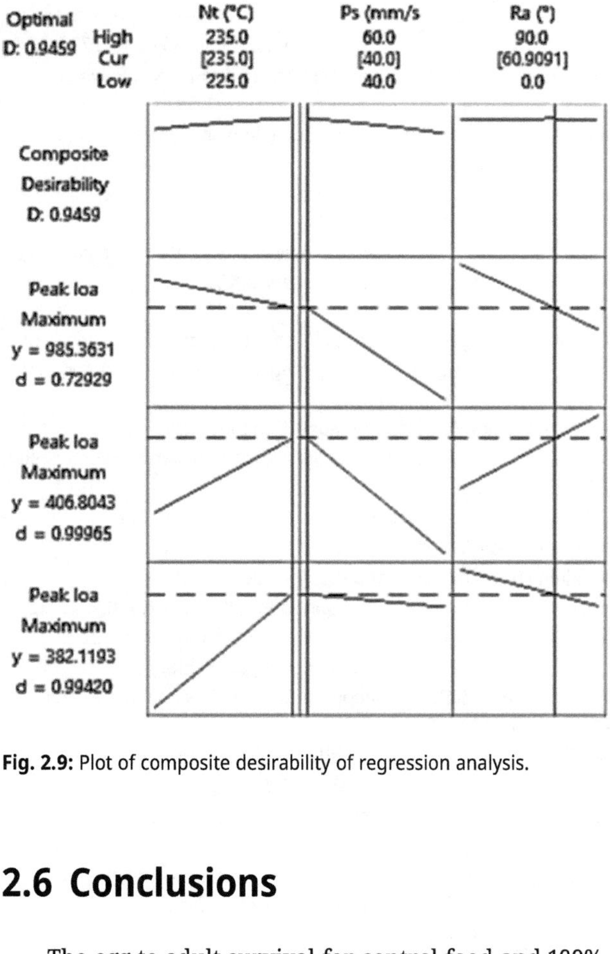

Fig. 2.9: Plot of composite desirability of regression analysis.

2.6 Conclusions

- The egg-to-adult survival for control food and 100% PVDF was observed to be the same in sample 2 (90% PVDF–10% HAp). Sample 3 (90% PVDF–8% HAp–2% CS) showed better egg-to-adult survival, but in sample 4 (90% PVDF–9% HAp–1% CS), the percentage of egg-to-adult survival was found to be less than in sample 3. The observation of the biocompatibility test of the PVDF composite shows that the best combination of HAp and CS in the PVDF matrix was found in sample 3, that is, 90% PVDF–8% HAp–2% CS for the egg-to-adult survival of *D. melanogaster*.

– The result suggests that from the multifactor optimization of printing process parameters of FFF were 235 °C NT, 40 mm/s Ps, 60° Ra suggested by the regression model, and 90° Ra suggested by a general linear model.

References

Andreu-Perez J, Leff DR, Ip HM, Yang GZ. From wearable sensors to smart implants-toward pervasive and personalized healthcare. IEEE Transactions on Biomedical Engineering. 2015 Apr 13;62(12):2750–62.

Chen FM, Liu X. Advancing biomaterials of human origin for tissue engineering. Progress in Polymer Science. 2016 Feb 1;53:86–168.

Husain M, Singh R, Pabla BS. A review on 3D printing of partially absorbable implants. Journal of the Institution of Engineers (India): Series C. 2023a Jul;23:1–20.

Husain M, Singh R, Pabla BS. On 3D printing of PVDF composite-based sensors for biomedical applications. National Academy Science Letters. 2023b Jun;22:1–6.

Husain M, Singh R, Pabla BS. On compressive properties of 3D printed polyvinylidene fluoride composite for implant applications. Journal of Thermoplastic Composite Materials. 2023c Sep;5:08927057231200005.

Husain M, Singh R, Pabla BS. On PVDF composite as partially absorbable smart implants. Proceedings of the Institution of Mechanical Engineers, Part H: Journal of Engineering in Medicine. 2023d Apr;237 (4):517–26.

Husain M, Singh R, Pabla BS. On digital twinning of fused filament fabrication for tensile properties of polyvinylidene fluoride composites-based functional prototypes. Journal of Thermoplastic Composite Materials. 2023e:08927057231211221.

Javaid M, Haleem A, Singh RP, Suman R. 3D printing applications for healthcare research and development. Global Health Journal. 2022 Nov 3.

Kadhim KT, Alsahlany AM, Wadi SM, Kadhum HT. An overview of patient's health status monitoring system based on the Internet of things (IoT). Wireless Personal Communications. 2020 Oct;114(3):2235–62.

Kumar R, Singh R, Husain M. Effect of chemical-assisted friction stir welding of thermoplastics. In: Materials Science and Materials Engineering. Elsevier, 2023a, ISBN 9780128035818.

Kumar V, Singh R, Kumar R, Ranjan N, Hussain M. ABS-PLA-Al composite for digital twinning of heritage buildings. Materials Letters. 2023b Sep 1;346:134536.

Ledet EH, Liddle B, Kradinova K, Harper S. Smart implants in orthopedic surgery, improving patient outcomes: A review. Innovation and Entrepreneurship in Health. 2018 Aug;29:41–51.

Morrison RJ, Kashlan KN, Flanagan CL, Wright JK, Green GE, Hollister SJ, Weatherwax KJ. Regulatory considerations in the design and manufacturing of implantable 3D-printed medical devices. Clinical and Translational Science. 2015 Oct;8(5):594–600.

Nasser F, Lynch I. Secreted protein eco-corona mediates uptake and impacts of polystyrene nanoparticles on Daphnia Magna. Journal of Proteomics. 2016 Mar 30;137:45–51.

Pradhan SR, Husain M, Singh R, Banwait SS. 8 A case study of 4D-imaging-assisted. In: 4D Imaging to 4D Printing: Biomedical Applications. 2022 Dec 30, 119.

Shao Z, Tao T, Xu H, Chen C, Lee IS, Chung S, Dong Z, Li W, Ma L, Bai H, Chen Q. Recent progress in biomaterials for heart valve replacement: Structure, function, and biomimetic design. View. 2021 Dec;2(6):20200142.

Singh R, Husain M, Mehra S. Multi-material 3D printing using DMLS. In: Materials Science and Materials Engineering. Elsevier, 2023a, ISBN 9780128035818.
Singh R, Husain M. Bone model studies for DMLS based 3D printed innovative dental implant. In: Materials Science and Materials Engineering. Elsevier, 2023b, ISBN 9780128035818.
Singh R, Husain M. Bone model studies for DMLS-based 3D printed intramedullary pin. In: Materials Science and Materials Engineering. Elsevier, 2023c, ISBN 9780128035818.
Singh R, Husain M. Orthopedic implant fabrication for canine using DMLS: A case study. In: Materials Science and Materials Engineering. Elsevier, 2023d, ISBN 9780128035818.
Weiland JD, Humayun MS. A biomimetic retinal stimulating array. IEEE Engineering in Medicine and Biology Magazine. 2005 Sep 26;24(5):14–21.

Minhaz Husain*, Rupinder Singh, Bahadur Singh Pabla

3 3D-printed self-energized energy storage device for biomedical applications

Abstract: In the past decade, a lot of work has been reported on polyvinylidene fluoride (PVDF) as an energy storage device (ESD), but hitherto little has been reported on self-energized ESD for biomedical applications. In this chapter, the piezoelectric qualities of β-phase PVDF are used as ESD because PVDF exists in a variety of crystalline phases, such as the alpha (α), beta (β), and gamma (γ) phases. Each of these stages has unique characteristics and uses. Due to its piezoelectric properties, which can aid in dissipating static charges, the phase of PVDF can be particularly intriguing in the context of ESD, as an example of how PVDF in the β-phase may be important for ESD. Piezoelectricity, or the ability to produce an electric charge in reaction to mechanical stress or deformation under compressive loading, is a property of PVDF in the β-phase. Numerous ESD applications, such as those involving piezoelectric sensors or actuators for ESD protection, can take advantage of this characteristic. The antistatic characteristics of composite materials containing PVDF–hydroxyapatite (HAp)–chitosan (CS) in the β-phase may be tailored. The finding suggests that the maximum β-phase was generated during a compressive load of 1,321 N on the PVDF–HAp–CS composite. The processing parameters for maximum generation of the β-phase of PVDF composite were 235 °C nozzle temperature, 60 mm/s printing speed, and 45° raster angle of fused filament fabrication process.

***Corresponding author: Minhaz Husain**, Department of Mechanical Engineering, National Institute of Technical Teachers Training and Research, Chandigarh, India, e-mail: minhazme2012@gmail.com
Rupinder Singh, Bahadur Singh Pabla, Department of Mechanical Engineering, National Institute of Technical Teachers Training and Research, Chandigarh, India

https://doi.org/10.1515/9783111081632-003

Graphical abstract:

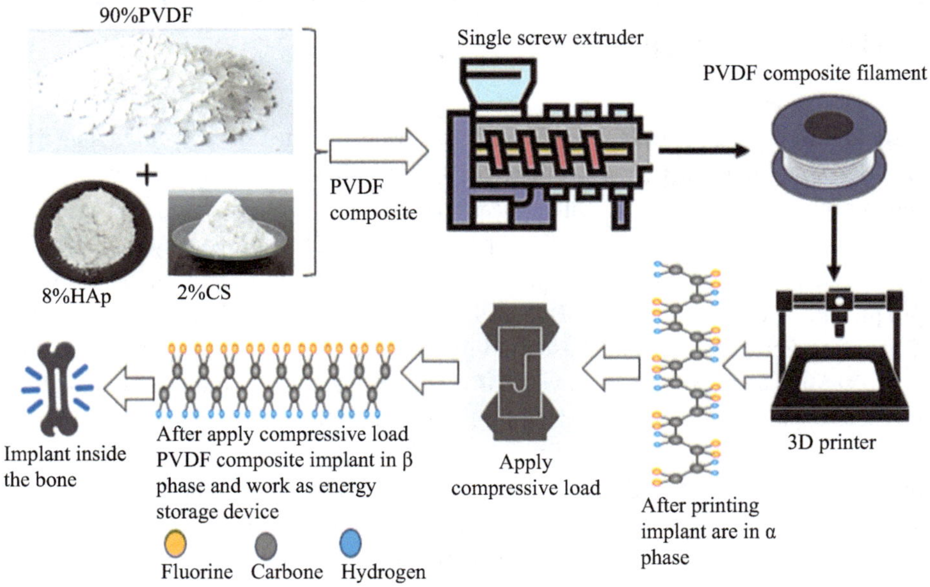

Keywords: 3D printing, self-energized energy storage, digital twin, biomedical implant

3.1 Introduction

The area of biomedical applications has undergone a revolution, thanks to three-dimensional (3D) printing, sometimes referred to as additive manufacturing (Ni et al. 2019). It has special benefits including the capacity to manufacture intricate and personalized structures with a high degree of precision and accuracy (Nouri et al. 2021). Personalized medicine, regenerative medicines, and patient care have all been advanced by the use of 3D printing in healthcare (Kumar et al. 2023). The technique opens up previously unimaginable possibilities for tissue engineering, medication delivery systems, surgical planning, and individually designed medical equipment. Overcoming difficulties with regulatory concerns, material biological compatibility, flexibility, and affordability will be critical as the sector develops (Kumar et al. 2023). In future, biomedical applications are predicted to be greatly influenced by 3D printing, thanks to continuing developments (Bozkurt et al. 2021). The use of 3D printing in medical fields shows how it might affect healthcare (Dodziuk 2016).

- Tissue modeling and restorative medicine (Arkhipov et al. 2017, Singh et al. 2023a): 3D printing makes it possible to create scaffolds with exact geometries and porosities that resemble the extracellular matrix of different tissues and organs. These scaffolds can aid in tissue regeneration by supporting tissues mechanically and directing cell development. A specialized kind of 3D printing called bioprinting enables the spatially controlled deposition of live cells to produce functioning tissues and organs.

- Medication delivery systems (Kesten et al. 1993, Singh et al. 2023b): 3D printing can help with controlled medication release systems and personalized therapeutics. Implants, microneedles, and microfluidic instruments are just a few examples of patient-specific medication delivery systems that are now possible. These technologies can improve patient compliance, lower adverse effects, and increase treatment effectiveness.
- Surgical management and healthcare education (Arden et al. 2021, Singh et al. 2023c): Using 3D printing and data from medical imaging, patient-specific anatomical models may be created. These models can be used by surgeons for preoperative planning, enhancing surgical results, and speeding up operations. In addition, 3D-printed models are useful teaching aids for educating medical staff and improving patient interaction.
- Orthodontics and prosthetics (Shroff et al. 1996, Singh et al. 2023d): 3D printing enables the creation of specialized prostheses and orthotics with better functioning, fit, and comfort. The technology makes it possible to create lightweight, inexpensive devices, boosting accessibility to assistive technology for those with physical limitations or limb loss.
- The invention of medical equipment (Lettl et al. 2006, Pradhan et al. 2022): 3D printing makes it possible to quickly prototype and produce sophisticated medical items, including implants, surgical tools, and anatomical models. It enables improved device design, patient-specific personalization, and cost-effective, efficient manufacture.
- Medication monitoring (Nash et al. 2011, Husain et al. 2023b): 3D printing may be used to create custom medication dosage forms and pharmaceuticals. It makes it possible to develop medications that are precisely suited to each patient's drug release profile. Additionally, the use of 3D-printed tissue models for medication screening and testing can eliminate the requirement for animal models and speed up the process of developing new drugs.

Due to their potential for sustaining numerous biomedical systems and equipment, self-energy storage device (SESD) has drawn a lot of interest in biomedical applications (Ma et al. 2017). The development of SESD for biomedical applications includes batteries, supercapacitors, and generating energy technologies. In addition, their particular uses are in point-of-care diagnostics, biosensors, and ubiquitous and implantable biomedical devices (Sendrowicz et al. 2022). The goal is to provide practitioners and researchers with a thorough grasp of the modern facilities in SESD technologies and their enormous potential for enhancing biomedical applications. Electrochemical procedures, electrode materials (such as lithium cobalt oxide and lithium iron phosphatase), electrolytes, and separators are all related to lithium-ion batteries; thin-film batteries, microbatteries, flexible and miniature batteries, and materials for flexible substrates; aqueous batteries, solid-state batteries, sodium-ion batteries, and other battery chemistries (Lakshmi et al. 2023).

Wearable and implanted medical devices, implantable pacemakers (Greenspon et al. 2011), implantable cardioverter–defibrillators (Kancharla et al. 2021), power needs, using energy for self-sustainability, and biocompatible packaging are all included in the integration of SESD in biomedical systems – systems for continuous glucose monitoring, wireless data transfer, and small-scale ESD (Kancharla et al. 2021), integrating SESD for accurate therapeutic administration and on-demand power supply with neurostimulators and medication delivery systems. Portable biosensors, self-sustaining sensing channels (Arsand et al. 2005), cost-effective data processing, and wireless transmission are all examples of biosensors and point-of-care diagnostics; paper-based diagnostic tools and the incorporation of energy storage tools for mobile and self-contained diagnostic systems; and miniaturized SESD with implantable biosensors for overtime and self-sufficient sensing applications. Energy density and device miniaturization improvements; research into novel materials, nanostructured electrodes, and advanced manufacturing processes; encapsulation strategies, electrode materials resistant to degradation, biocompatibility, and long-term stability in design; creation of wireless power transfer schemes and intelligent power management algorithms (Timurdogan et al. 2017).

Polyvinylidene fluoride (PVDF) transformation from α- to β-phase (Husain et al. 2023a) is a key occurrence that affects the material's properties, notably its piezoelectric and magnetic traits (Gunduz et al. 2014). The nonpolarized (α-stage) and the polarized (β-stage) are the two different crystal phases of PVDF (Kumar et al. 2019). At ambient temperature, the β-phase possesses improved piezoelectric and ferroelectric capabilities while the α-phase is the stable form. PVDF can undergo an α- to β-stage conversion utilizing several techniques, such as mechanical loading (in this case, compressive loading), the use of heat Diffrential Scanning Calorimetry (DSC) (Martins et al. 2009), and electric poling (Benz and Euler 2003). The choice is based on the intended use and particular needs, and each approach has pros and cons.

- Mechanical loading: The PVDF sample can undergo an α- to β-phase transformation during mechanical loading or uniaxial strain is applied. Loading straightens the polymer links and encourages crystalline structure rearrangement, which favors the creation of the β-phase. Loading intensity and strain rate have a big impact on phase conversion and subsequent characteristics (Husain et al. 2023c).
- Heating (DSC): Another technique used to transition PVDF from the α- to β-phase is thermal treatment (DSC). PVDF may be crystallized in the β-phase by slowly cooling it after being heated over its melting point (about 170–180 °C). To maximize the β-phase content and achieve the desired phase transition, it is necessary to control the annealing temperature, time, and cooling rate (Husain et al. 2023d).

- Electric poling: In other methods used for the conversion of α- to β-phase transition in PVDF films, an intense electric field is applied during electric poling. The polar β-phase rearrangement is made easier by the electric field's alignment of the poles in the thermoplastic's chains. The rate of β-phase conversion depends on the strength and period of the magnetic field, the temperature, and the poling process.

Changes in material characteristics, such as improved piezoelectric, enhanced mechanical properties, better ferroelectric polarization, and greater charge storage capability (ε_r), occur together with the α- to β-phase transformation in PVDF. The α- to β-phase conversion applications such as sensors that use piezoelectric, and 4D properties are particularly interested in this transition. The material tends to return to its original phase as time goes on; therefore, it can be difficult to keep PVDF in the β-phase. Several techniques, including cross-linking, nanoestablishing, and combining with other plastics, can be used to maintain the β-phase.

3.2 Background

Bibliometric analysis was done with the Web of Science database for the last 20 years. The keyword "energy storage device (ESD) of thermoplastics" was explored on the Web of Science site and found 79 results. These results were analyzed with open VOSviewer software for considering 5 minimum number of occurrences, total terms were found 2,830, and 59 terms meet the thresholds. For each of the 59 terms, a relevance score will be calculated as given in Tab. 3.1. Based on this score, the most relevant terms will be selected. The default choice is to select 60% of most relevant terms. The number of terms selected 35 for the network diagram is shown in Fig. 3.1. In Fig. 3.2, the reported work is highlighted and other details for the keyword "energy storage device of thermoplastics" are shown in Fig. 3.3: (a) author details, (b) publication years, (c) publications houses, and (d) countries or regions for the keyword.

Similarly, another keyword "self-energized energy storage device (SEESD) for biomedical application" was explored in the Web of Science database for the last 20 years and found only two results. These results were analyzed with open VOSviewer software for considering 1 minimum number of occurrence; a total of 125 terms were found, and these terms met the thresholds. For each of the 125 terms, a relevance score will be calculated as given in Tab. 3.2. Based on this score, the most relevant terms will be selected. The default choice is to select 60% most relevant terms. The number of terms (75) to be selected for the network diagram is shown in Fig. 3.4(a), and the research gap for the node of the medical device is shown in Fig 3.4(b).

Figures 3.1–3.4 depict the results from the bibliometric analysis. There is no work reported on SEESD for biomedical applications.

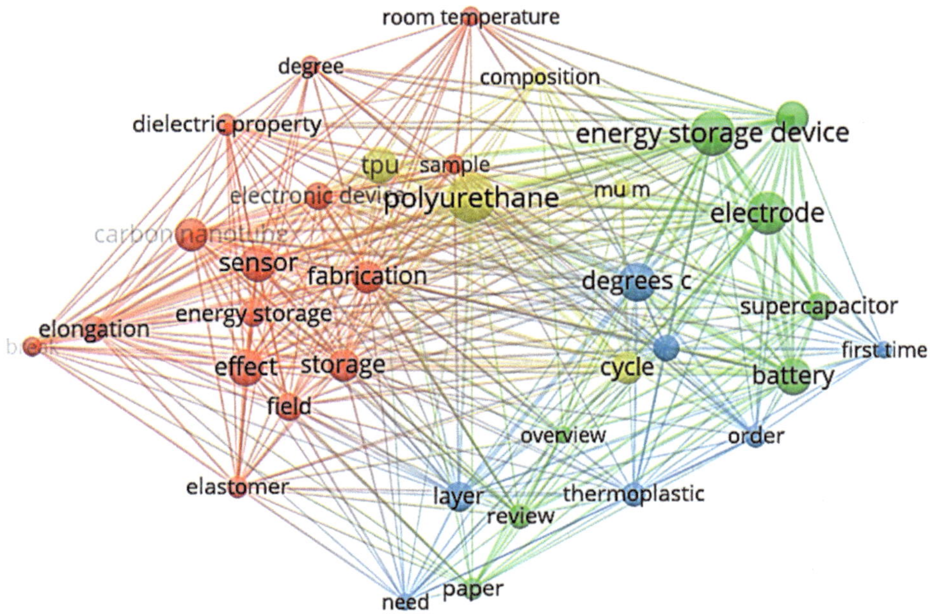

Fig. 3.1: Bibliometric network diagram for ESD of thermoplastics.

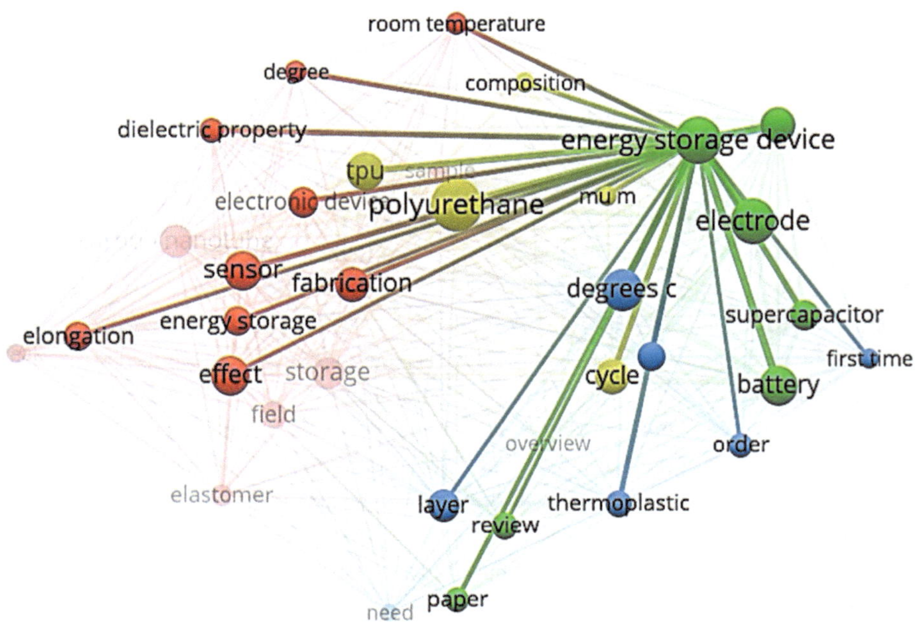

Fig. 3.2: Research gap for ESD.

Tab. 3.1: Occurrence table for the keyword ESD of thermoplastics.

Id	Terms	Occurrences	Relevance scores
1	Approach	9	0.664
2	Battery	15	1.1586
3	Break	6	3.0169
4	Carbon nanotube	13	0.6649
5	Composition	5	0.5683
6	Cycle	12	0.3575
7	Degree	6	0.563
8	Degrees c	16	0.6505
9	Dielectric property	7	0.7014
10	Effect	15	0.861
11	Elastomer	7	1.5433
12	Electrode	18	0.946
13	Electrolyte	13	1.3883
14	Electronic device	10	0.5633
15	Elongation	9	1.9959
16	Energy storage	9	0.7871
17	Energy storage device	20	1.0035
18	Fabrication	12	0.6099
19	Field	10	1.2285
20	First time	5	2.406
21	Layer	11	0.6924
22	Mum	5	0.5513
23	Need	5	1.0711
24	Order	7	1.3321
25	Overview	5	1.7521
26	Paper	7	0.644
27	Polyurethane	23	0.2142
28	Review	8	1.2268
29	Room temperature	6	0.5686
30	Sample	6	0.5962
31	Sensor	15	0.7808
32	Storage	12	0.6324
33	Supercapacitor	10	1.9653
34	Thermoplastic	8	0.9977
35	TPU	14	0.2973

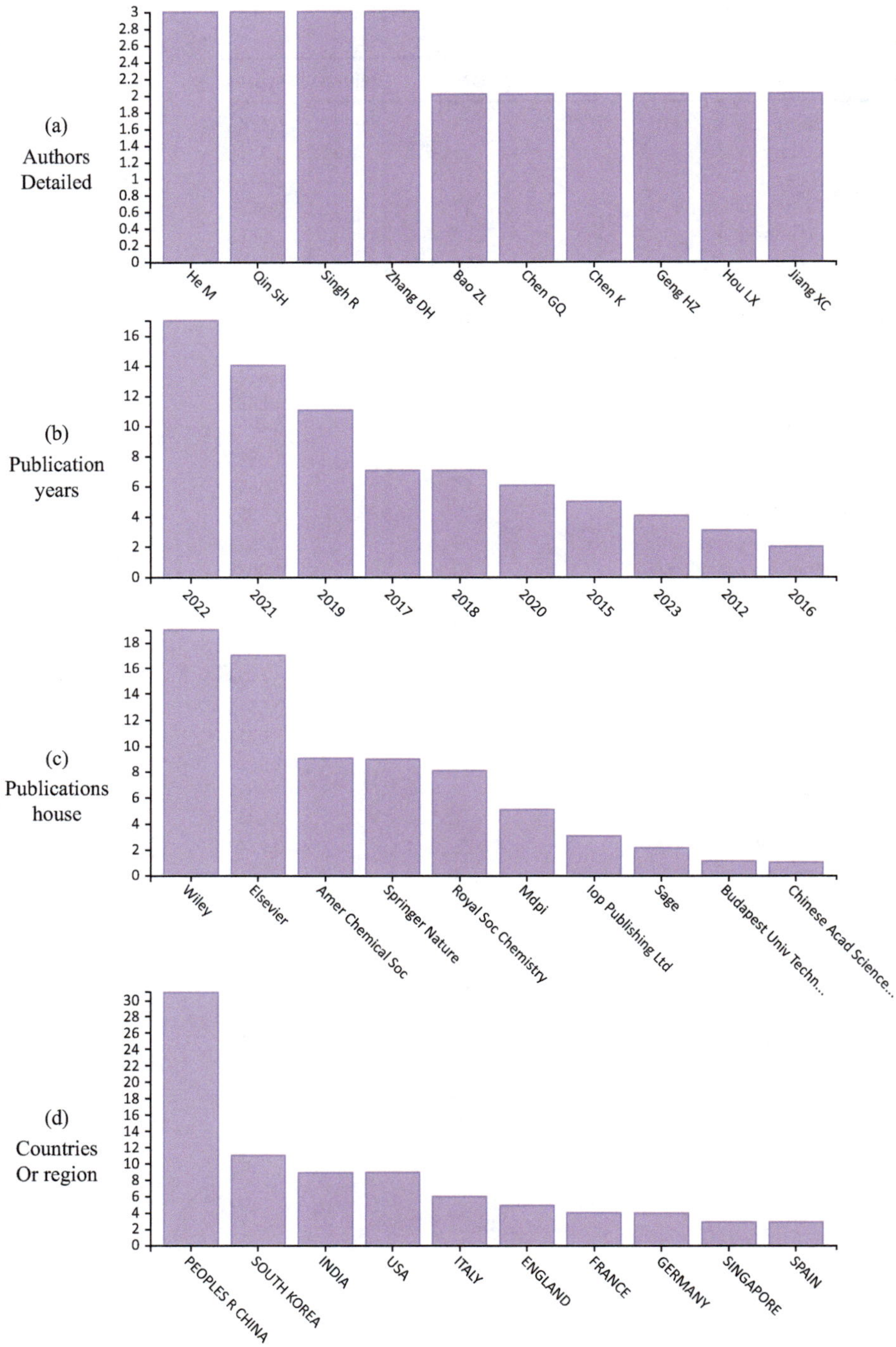

Fig. 3.3: (a) Author details, (b) publication years, (c) publication houses, and (d) countries or regions for the keyword "energy storage device of thermoplastics."

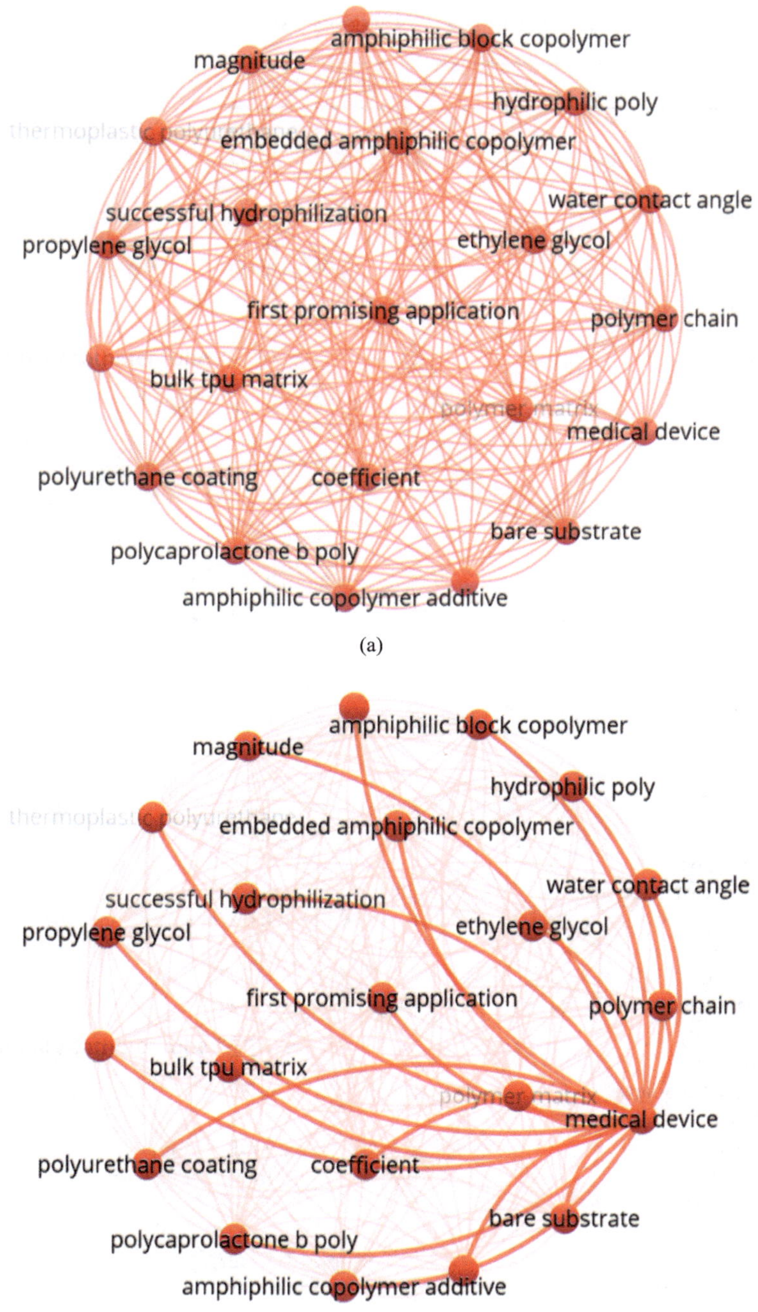

Fig. 3.4: (a) Network analysis for the keyword "SEESD for biomedical application" and (b) research gap for the node of medical device.

Tab. 3.2: Occurrence table for the keyword "SEESD for biomedical application."

Id	Terms	Occurrences	Relevance scores
1	Amphiphilic block copolymer	1	0.799
2	Amphiphilic copolymer additive	1	0.799
3	Bare substrate	1	0.799
4	Bulk TPU matrix	1	0.799
5	Coefficient	1	0.799
6	Embedded amphiphilic copolymer	1	0.799
7	Ethylene glycol	1	0.799
8	First promising application	1	0.799
9	Hydrophilic polymers	1	0.799
10	Hydrophilization	1	0.799
11	Magnitude	1	0.799
12	Medical device	1	0.799
13	Open-cell foam	1	1.2327
14	Organic compound	1	1.2327
15	Other technological areas	1	1.2327
16	Piezoresistive sensor	1	1.2327
17	Polycaprolactone	1	0.799
18	Polymer chain	1	0.799
19	Polymer matrix	1	0.799
20	Polymeric foam	1	1.2327
21	Polypropylene	1	1.2327
22	Polystyrene foam	1	1.2327
23	Polyurethane coating	1	0.799
24	Practical challenge	1	1.2327
25	Propylene glycol	1	0.799
26	Special emphasis	1	1.2327
27	Successful hydrophilization	1	0.799
28	Surface-coated polymeric foams	1	1.2327
29	Surface modification	1	1.2327
30	Surface property	1	1.2327
31	Surface treatment	1	1.2327
32	Sustainable technological solution	1	1.2327
33	Thermal conductivity	1	1.2327
34	Thermoplastic foam	1	1.2327
35	Thermoplastic polyurethane	1	0.799
36	Urinary catheter	1	0.799
37	Urinary catheter tube	1	0.799
38	Various polymeric foams	1	1.2327
39	Water contact angle	1	0.799
40	Water remediation	1	1.2327
41	Wide spectrum	1	1.2327

3.3 Materials and methods

PVDF granules, hydroxyapatite (HAp), and chitosan (CS) were selected by wt% and generated a filament through a single screw extruder having a diameter of 1.75 ± 0.05 mm. This filament was used to print the compression sample with the help of an open-source printer (fused filament fabrication, FFF) to ascertain the peak load (PL) of the compressive sample test in the universal tensile machine. The adopted methodology is shown in Fig. 3.5.

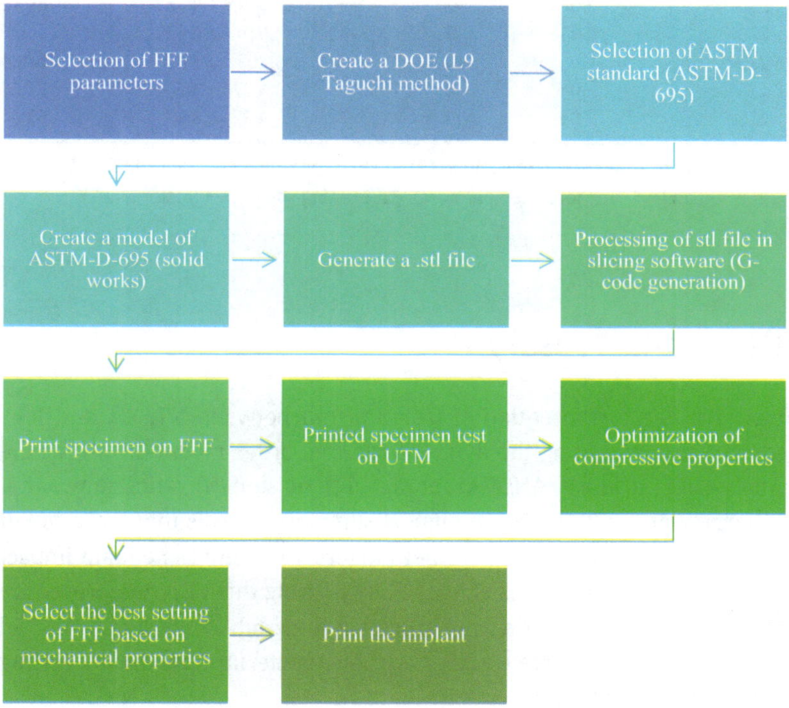

Fig. 3.5: Adopted methodology for this chapter.

3.4 Experimentation

A design of the experiment (DOE) was created using three factors (nozzle temperature (NT), printing speed (PS), and raster angle (RA)), and the three levels are given in Tab. 3.3. DOE analysis using the L9 Taguchi optimization technique is given in Tab. 3.3.

Tab. 3.3: Input, output, and S/N ratios of the parameters.

S. no.	Input characteristics			Output characteristics	S/N ratios
	NT (°C)	PS (mm/s)	RA (°)	PL (N)	
R1	225	40	0	563.4 ± 8.76	55.01
R2	225	50	45	500.6 ± 6.54	53.98
R3	225	60	90	716.4 ± 4.87	57.10
R4	230	40	45	904.2 ± 9.23	59.12
R5	230	50	90	1,019.4 ± 8.78	60.16
R6	230	60	0	1,019.4 ± 9.65	60.16
R7	235	40	90	1,295.5 ± 8.45	62.24
R8	235	50	0	1,270 ± 9.56	62.07
R9	235	60	45	1,321 ± 8.69	62.41

Note: The five repetitions for flexural testing of samples (R1 to R9) as per the modified ASTM-D-695 standard were performed on UTT, NT (nozzle temperature), PS (printing speed), RA (raster angle), and PL (peak load).

3.5 Results and discussion

Table 3.4 depicts the regression equation (relationship between NT, PS, and RA), which was required for generating the dynamic reduced-order model (ROM) for the digital twin. Analysis of variance (ANOVA) for signal-to-noise (S/N) ratios shows that the significant parameter for PL is NT. The model summary reveals that the value of R^2 is greater than 95%. When the procedure parameters are altered, the main impact on the trend of PL changes. The PL was enhanced by altering the NT in the procedure parameters. This is because high levels were crucial in the establishment of deposited strata. Increasing the NT could have enabled a proper material flow during setting up, leading to greater layer adhesion.

Real-time data and the development of dynamic ROM are the two main requirements for digital twinning (Kumar et al. 2023) of the FFF process for compression testing of PVDF composite-based functional prototypes. For real-time data, it is advised to use a sensor based on the concept of a microstrip patch antenna (MPA) (Husain et al. 2023b). The dielectric property of a typical MPA-based sensor is observed using a vector network analyzer (VNA) in terms of the dielectric constant (ε_r). The calibration of the VNA output in the industrial, scientific, and medical band for accessing this data in Bluetooth/Wi-Fi range using a smartphone as an Internet of things-based solution has received significant attention in recent years. The objective is to quantify the change in ε_r under compressive loading using FFF in-process 3D printing. In other words, the printed sample based on FFF changes the value of ε_r before as well as following the compression test. Similar to this, the sensor's ε_r value will change throughout 3D printing (when the sample is created using different processing parameters).

Tab. 3.4: Regression equation, ANOVA for S/N ratios, model summary, coefficient, main effect, and residual plots for PL.

Regression equation	**PL (N) = – 15, 465 + 70.20 NT (°C) + 4.89 PS (mm/s) + 0.661 RA (°)**							
ANOVA for S/N ratios	Source	DF	Seq. SS	Contribution	Adj. SS	Adj. MS	*F*-value	*P*-value
	Regression	3	758,963	97.71%	758,963	252,988	71.19	0.000
	NT (°C)	1	739,276	95.18%	739,276	739,276	208.03	0.000
	PS (mm/s)	1	14,377	1.85%	14,377	14,377	4.05	0.100
	RA (°)	1	5,310	0.68%	5,310	5,310	1.49	0.276
	Error	5	17,769	2.29%	17,769	3,554		
	Total	8	776,732	100.00%				
Model summary	*S*	R^2	R^2 (adj.)	PRESS	R^2 (pred.)	AICc	BIC	
	59.6131	97.71%	96.34%	49,634.9	93.61%	123.83	104.82	
Coefficients	Term	Coefficient	SE coefficient	95% CI	*T*-value	*P*-value	VIF	
	Constant	−15,465	1,127	(−18,360, −12,569)	−13.73	0.000		
	NT (°C)	70.20	4.87	(57.69, 82.72)	14.42	0.000	1.00	
	PS (mm/s)	4.89	2.43	(−1.36, 11.15)	2.01	0.100	1.00	
	RA (°)	0.661	0.541	(−0.729, 2.051)	1.22	0.276	1.00	

(continued)

Tab. 3.4 (continued)

Regression equation	**PL (N) = – 15, 465 + 70.20 NT (°C) + 4.89 PS (mm/s) + 0.661 RA (°)**
Main effect plots for S/N ratios	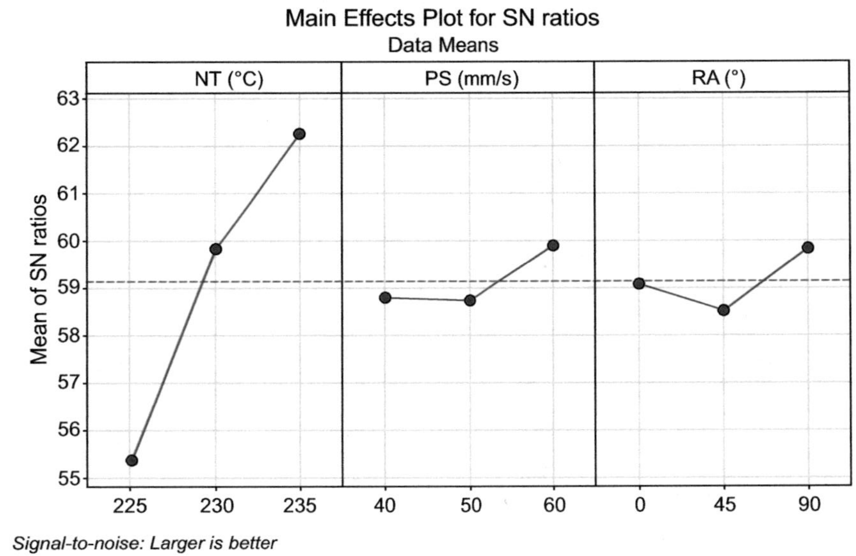

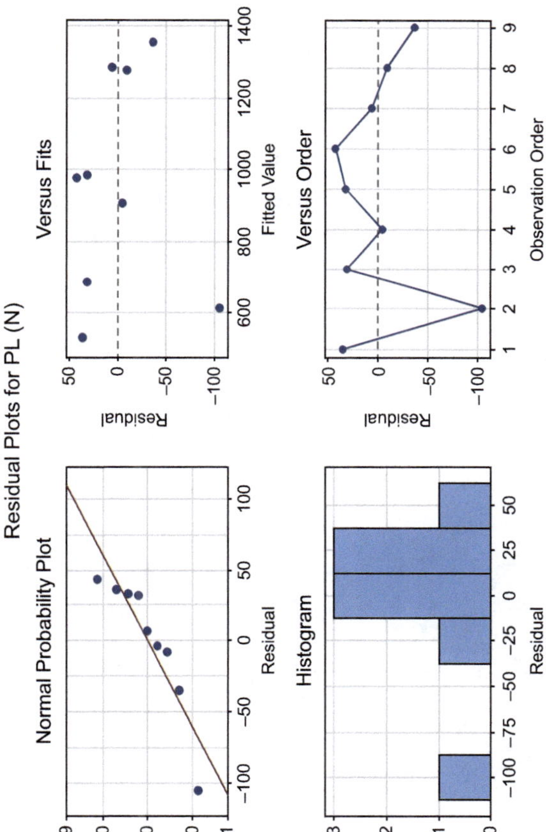

Residual plots for S/N ratios

For instance, when producing high infill density as opposed to low infill density, the ε_r value will change. Similar to this, the effects of NT, PS, and other input variables will be felt. This helps to generate real-time data while printing and simulates the change in ε_r during destructive analysis (compressive test). Therefore, one may predict the physical features online during printing themselves (by using the dynamic ROM developed based on earlier experimental data) before going on to the destructive testing step. The schematic of the recommended digital twin is shown in Fig. 3.6.

The PVDF is renowned for its piezoelectric qualities and the capacity to exist in a variety of crystalline phases, such as the alpha (α), beta (β), and gamma (γ) phases. Each of these stages has unique characteristics and uses. Due to its piezoelectric properties, which can aid in dissipating static charges, the phase of PVDF can be particularly intriguing in the context of ESD, as an example of how PVDF in the β-phase may be important for ESD. Piezoelectricity, or the ability to produce an electric charge in reaction to mechanical stress or deformation, is a property of PVDF in the β-phase. Numerous ESD applications, such as those involving piezoelectric sensors or actuators for ESD protection, can take advantage of this characteristic.

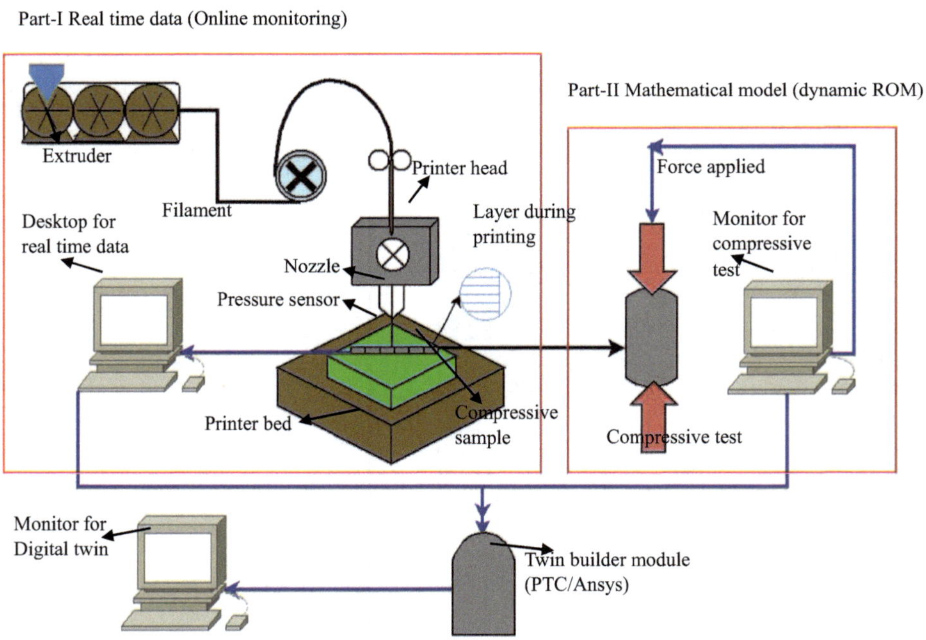

Fig. 3.6: Functional prototype of a digital twin for compressive test.

The antistatic characteristics of composite materials containing PVDF in the β-phase may be tailored. PVDF may be used to make composite materials, including polymer matrixes, that can help disperse static charges and lessen the likelihood of ESD incidents. Conducive coatings, films, or textiles made of PVDF-based composites with the phase can be employed to create ESD protection devices. To prevent static discharge, these materials can be applied to surfaces like electrical components or packaging. Sensors that can pick up changes in pressure, vibration, or electric charge employ PVDF-based materials from the phase. In ESD-sensitive areas, where monitoring and regulation of static charge levels are essential, these sensors can be useful.

It is crucial to remember that the characteristics and efficiency of PVDF-based composites for ESD applications might change depending on the precise formulation, processing, and intended usage. The electrical and mechanical characteristics of PVDF materials may also exist in a variety of phases, and it is crucial to regulate the crystalline phase if you want to adjust the material's performance for the intended use in an ESD application.

To guarantee that the PVDF composite material is developed and configured to satisfy the unique needs of the application, it is recommended to engage materials scientists and engineers who specialize in polymer and composite materials. To confirm that the composite is indeed effective at avoiding electrostatic discharge, it should be subjected to appropriate testing and assessment of its ESD performance.

3.6 Conclusions

The application of ESD in the biomedical field, such as those involving piezoelectric sensors or actuators for ESD protection, can take advantage of the PVDF composite characteristic. The antistatic characteristics of composite materials containing PVDF–HAp–CS in the β-phase may be tailored using various reinforcements in the PVDF matrix. The finding suggests that there is maximum β-phase generation during compressive load of 1,321 N on PVDF–HAp–CS composite. The processing parameters for maximum generation of the β-phase of PVDF composite were 235 °C NT, 60 mm/s PS, and 45° raster angle of the FFF process.

References

Arden NK, Perry TA, Bannuru RR, Bruyère O, Cooper C, Haugen IK, Hochberg MC, McAlindon TE, Mobasheri A, Reginster JY. Non-surgical management of knee osteoarthritis: Comparison of ESCEO and OARSI 2019 guidelines. Nature Reviews Rheumatology. 2021 Jan;17(1):59–66.

Arkhipov M, Leskov A, Golovin V, Gercik Y, Kocherevskaya L. Prospects of robotics development for restorative medicine. Advances in Robot Design and Intelligent Control: Proceedings of the 25th Conference on Robotics in Alpe-Adria-Danube Region (RAAD16) 2017 (pp. 499–506). Springer International Publishing.

Arsand E, Walseth OA, Andersson N, Fernando R, Granberg O, Bellika JG, Hartvigsen G. Using blood glucose data as an indicator for epidemic disease outbreaks. Studies in Health Technology and Informatics. 2005 Jan 1;116:217–22.

Benz M, Euler WB. Determination of the crystalline phases of poly(vinylidene fluoride) under different preparation conditions using differential scanning calorimetry and infrared spectroscopy. Journal of Applied Polymer Science. 2003 Jul 25;89(4):1093–100.
Bozkurt Y, Karayel E. 3D printing technology; methods, biomedical applications, future opportunities and trends. Journal of Materials Research and Technology. 2021 Sep 1;14:1430–50.
Dodziuk H. Applications of 3D printing in healthcare. Kardiochirurgia I Torakochirurgia Polska/Polish Journal of Thoracic and Cardiovascular Surgery. 2016 Sep 30;13(3):283–93.
Greenspon AJ, Patel JD, Lau E, Ochoa JA, Frisch DR, Ho RT, Pavri BB, Kurtz SM. 16-year trends in the infection burden for pacemakers and implantable cardioverter-defibrillators in the United States: 1993 to 2008. Journal of the American College of Cardiology. 2011 Aug 30;58(10):1001–06.
Gunduz D, Stamatiou K, Michelusi N, Zorzi M. Designing intelligent energy harvesting communication systems. IEEE Communications Magazine. 2014 Jan 16;52(1):210–16.
Husain M, Singh R, Pabla BS. A review on 3D printing of partially absorbable implants. Journal of the Institution of Engineers (India): Series C. 2023a Jul;23:1–20.
Husain M, Singh R, Pabla BS. On 3D printing of PVDF composite-based sensors for biomedical applications. National Academy Science Letters. 2023b Jun;22:1–6.
Husain M, Singh R, Pabla BS. On compressive properties of 3D printed polyvinylidene fluoride composite for implant applications. Journal of Thermoplastic Composite Materials. 2023c Sep;5:08927057231200005.
Husain M, Singh R, Pabla BS. On PVDF composite as partially absorbable smart implants. Proceedings of the Institution of Mechanical Engineers, Part H: Journal of Engineering in Medicine. 2023d Apr;237 (4):517–26.
Kancharla K, Hayes DL, Asirvatham SJ, Friedman PA. Indications for pacemakers, implantable cardioverter-defibrillators, and cardiac resynchronization therapy: Identifying patients who benefit from cardiac rhythm devices. Cardiac Pacing, Defibrillation and Resynchronization: A Clinical Approach. 2021 Apr;19:86–124.
Kesten S, Zive K, Chapman KR. Pharmacist knowledge and ability to use inhaled medication delivery systems. Chest. 1993 Dec 1;104(6):1737–42.
Kumar R, Singh R, Husain M. Effect of chemical-assisted friction stir welding of thermoplastics. In: Materials Science and Materials Engineering. Elsevier, 2023, ISBN 9780128035818.
Kumar RS, Sarathi T, Venkataraman KK, Bhattacharyya A. Enhanced piezoelectric properties of polyvinylidene fluoride nanofibers using carbon nanofiber and electrical poling. Materials Letters. 2019 Nov 15;255:126515.
Kumar V, Singh R, Kumar R, Ranjan N, Hussain M. ABS-PLA-Al composite for digital twinning of heritage buildings. Materials Letters. 2023 Sep 1;346:134536.
Lakshmi D, Palaniyandy N, Bohm S, Mamba BB. Understanding the mechanisms of mixed-ion cathode materials for aqueous and non-aqueous lithium/sodium-ion batteries–A review. Current Opinion in Electrochemistry. 2023 Jan;21:101217.
Lettl C, Herstatt C, Gemuenden HG. Users' contributions to radical innovation: Evidence from four cases in the field of medical equipment technology. R&D Management. 2006 Jun;36(3):251–72.
Ma M, Gan Z, Xu Z. Ion structure near a core-shell dielectric nanoparticle. Physical Review Letters. 2017 Feb 16;118(7):076102.
Martins P, Nunes JS, Hungerford G, Miranda D, Ferreira A, Sencadas V, Lanceros-Méndez S. Local variation of the dielectric properties of poly(vinylidene fluoride) during the α-to β-phase transformation. Physics Letters A. 2009 Jan 5;373(2):177–80.
Nash M. Improving mental health service users' physical health through medication monitoring: A literature review. Journal of Nursing Management. 2011 Apr;19(3):360–65.
Ni J, Ling H, Zhang S, Wang Z, Peng Z, Benyshek C, Zan R, Miri AK, Li Z, Zhang X, Lee J. Three-dimensional printing of metals for biomedical applications. Materials Today Bio. 2019 Jun 1;3:100024.

Nouri A, Shirvan AR, Li Y, Wen C. Additive manufacturing of metallic and polymeric load-bearing biomaterials using laser powder bed fusion: A review. Journal of Materials Science & Technology. 2021 Dec 20;94:196–215.

Pradhan SR, Husain M, Singh R, Banwait SS. 8 A case study of 4D-imaging-assisted. 4D imaging to 4D printing. In: Biomedical Applications. 2022 Dec 30;119.

Sendrowicz A, Myhre AO, Yasnikov IS, Vinogradov A. Stored and dissipated energy of plastic deformation revisited from the viewpoint of dislocation kinetics modeling approach. Acta Materialia. 2022 Sep 15;237:118190.

Shroff B, Siegel SM, Feldman S, Siegel SC. Combined orthodontic and prosthetic therapy: Special considerations. Dental Clinics of North America. 1996 Oct 1;40(4):911–43.

Singh R, Husain M, Mehra S. Multi-material 3D printing using DMLS. In: Materials Science and Materials Engineering. Elsevier, Vol. 2023a, ISBN 9780128035818.

Singh R, Husain M. Bone model studies for DMLS based 3D printed innovative dental implant. In: Materials Science and Materials Engineering. Elsevier, 2023b, ISBN 9780128035818.

Singh R, Husain M. Bone model studies for DMLS-based 3D printed intramedullary pin. In: Materials Science and Materials Engineering. Elsevier, 2023c, ISBN 9780128035818.

Singh R, Husain M. Orthopedic implant fabrication for canine using DMLS: A case study. In: Materials Science and Materials Engineering. Elsevier, 2023d, ISBN 9780128035818.

Timurdogan E, Alaca BE, Kavakli IH, Urey H. MEMS biosensor for detection of Hepatitis A and C viruses in serum. Biosensors and Bioelectronics. 2011 Oct 15;28(1):189–94.

Vinay Kumar*, Nishant Ranjan, Rupinder Singh,
Inderpreet Singh Ahuja

4 4D printing of smart thermoplastic composites for online health monitoring

Abstract: The past two decades have expressed very advanced industrial utilities of smart thermoplastic composites for pressure sensing, aerospace, welding of dissimilar polymers, structural monitoring, 3D printing, and so on. Nevertheless, very little work has been reported on 4D printing of smart thermoplastic composites for sensing features based on online health monitoring (OHM) applications. This work highlights some of the investigations performed on dielectric properties of polyvinylidene fluoride (PVDF) based on various smart thermoplastic composites for OHM applications in structural and nonstructural engineering. The experimentations performed on a self-actuating smart composite matrix of PVDF have been observed as an innovative solution for the preparation of smart repair solutions for nonstructural cracks in heritage structures. To date, the 4D printing-based properties of such compositions are the least reported for smart sensor fabrication with OHM capabilities. As an extension to the previously reported studies on PVDF composite matrix-based smart customizable solutions (for the repair of heritage buildings), this study outlines the experimental trials performed on the effect of dielectric properties of smart thermoplastic composites for OHM of heritage structures. The outcomes of the work show that dielectric properties of the compositions (governed by rheological properties) have a significant effect on mechanical, thermal, morphological, and shape memory properties for OHM of repaired heritage structures. The multifactor optimization has been performed to ascertain the acceptable composition of polymer composites for efficient dielectric property-based sensor fabrication for health monitoring applications.

Acknowledgments: The authors are thankful for the financial support from the Department of Science and Technology (DST, Government of India) under the heritage preservation research initiative (file no. DST/TDT/SHRI-35/2018).

***Corresponding author: Vinay Kumar**, University Centre of Research and Development, Chandigarh University, Mohali, Punjab, India, e-mail: vinaykumar6192@gmail.com
Nishant Ranjan, University Centre of Research and Development, Chandigarh University, Mohali, Punjab, India
Rupinder Singh, Department of Mechanical Engineering, National Institute of Technical Teachers Training and Research, Chandigarh, India
Inderpreet Singh Ahuja, Department of Mechanical Engineering, Punjabi University, Patiala, Punjab, India

https://doi.org/10.1515/9783111081632-004

Graphical abstract:

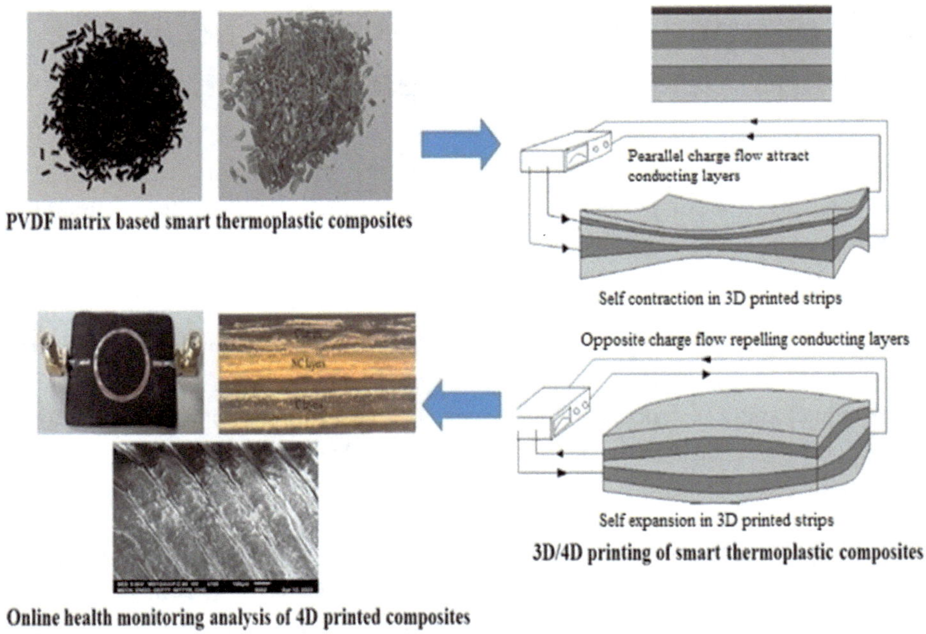

Keywords: PVDF thermoplastic composites, 4D printing, heritage structures, repair solutions, online health monitoring

4.1 Introduction

A significant work has been reported by researchers in the past two decades on smart thermoplastic composite-based materials for innovative consumer applications. The study of designs and mechanics of smart composite has outlined the applications of thermoplastic-based composites for a variety of consumer applications, including sensor fabrication, and biomedical applications (Elhajjar et al. 2013). Some investigations performed on smart composite materials have reported that controlled anisotropic properties of composites play a significant role in imparting shape recovery properties to the base polymer matrix for the desired utility (Varga et al. 2005). The design and development of metal-based, natural, and synthetic biomaterial-based composites and thermoplastic-based composites also have been reported by some researchers for the fabrication of smart implants, tissue regeneration, and 4D printing-based customizable solutions for heritage structures (Asanuma 2000, Perez et al. 2013, Kumar et al. 2023a). Such literature reported on smart composites in the past 50 years to recent times indicates that smart composites play a significant role in the manufacturing of advanced consumer products. The Scopus database indicated that a gradual increment has been observed in studies reported on various smart composites for novel manufacturing practices.

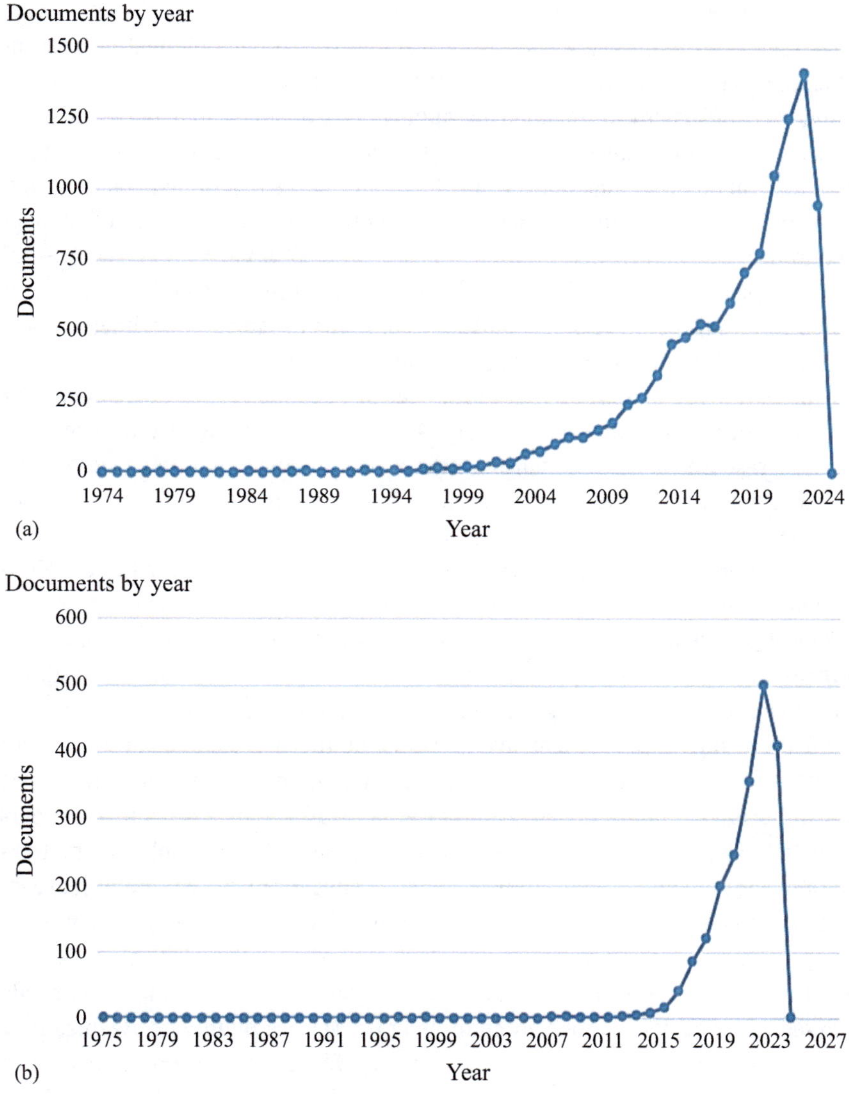

Fig. 4.1: Year-wise distribution of documents published in the past 50 years on smart composites for (a) health monitoring and (b) 4D printing.

As per the Scopus database, 10,646 documents have been reported in the past 50 years on the investigations performed by various researchers on health monitoring applications of composites, whereas 2014 documents have been reported in the recent past on 4D printing applications of smart composites. Figure 4.1a shows the observations for the year-wise distribution of documents published in the past 50 years on smart composites for health monitoring and Fig. 4.1b shows the trends obtained in publication of documents on 4D printing of smart composite materials.

Beyond the nonstructural engineering applications of smart thermoplastic composites, some researchers have reported cement-based piezoelectric ceramic composites for structural engineering applications (Dong and Li 2005). Graphene and carbon nanotube-reinforced composites also have been reported for electrostriction and magnetostriction-based structural requirements (Islam et al. 2022, Varga et al. 2005). Similar to such composites, in-house fabricated smart thermoplastic composites have been investigated for 3D/4D printing of customizable solutions for cultural heritage conservation (Kumar et al. 2022a). The wide area of application of smart thermoplastic composites justifies the publication of a large number of documents in this area over the past (as shown in Fig. 4.1a and b). Also, Fig. 4.2a and b shows the rise in year-wise publication of research articles by various sources on advanced manufacturing practices and condition monitoring research areas based on applications of smart thermoplastic composites.

The experimental investigations reported on smart composites for the development of smart optical sensors, piezoelectric actuators, and thermal strain sensors in the past 20 years have indicated the wide role of thermoplastic composites in continuously improving manufacturing practices focused on advanced consumer products (Tauchert et al. 2000, Lau 2003). It has been reported that the structural and nonstructural applications of stimulus-responsive and tunable thermoplastic composites may be used for optical data transfer as well as digital twin building of heritage structures and cultural heritage (Pucci et al. 2011, Kumar et al. 2023b). Figure 4.3 shows the percentage distribution of various types of documents published on polymer matrix-based composites indicating the usefulness of composites for different research areas.

One of the major reasons for the increasing number of documents on polymer-based composites may also lie in the environmental issues related to the consumption of thermoplastic goods. With time, the dependence of mankind on plastic-based goods has increased significantly by not only improving the service quality of organizations but also affecting the environment drastically. The fabrication of novel industrial products from recycled polymer-based composites by conventional and additive manufacturing processes contributed significantly to improving the environmental situations but also paved a new way for the preservation of a variety of artifacts and archaeological treasures around the globe (Umar et al. 2021, Ikram et al. 2022, Kumar et al. 2023c). Since the development of composite structure-based smart materials may be utilized effectively for solving the plastic-related environmental issues, the studies performed on plastic-based composites have reported an increase in the recent past for the fabrication

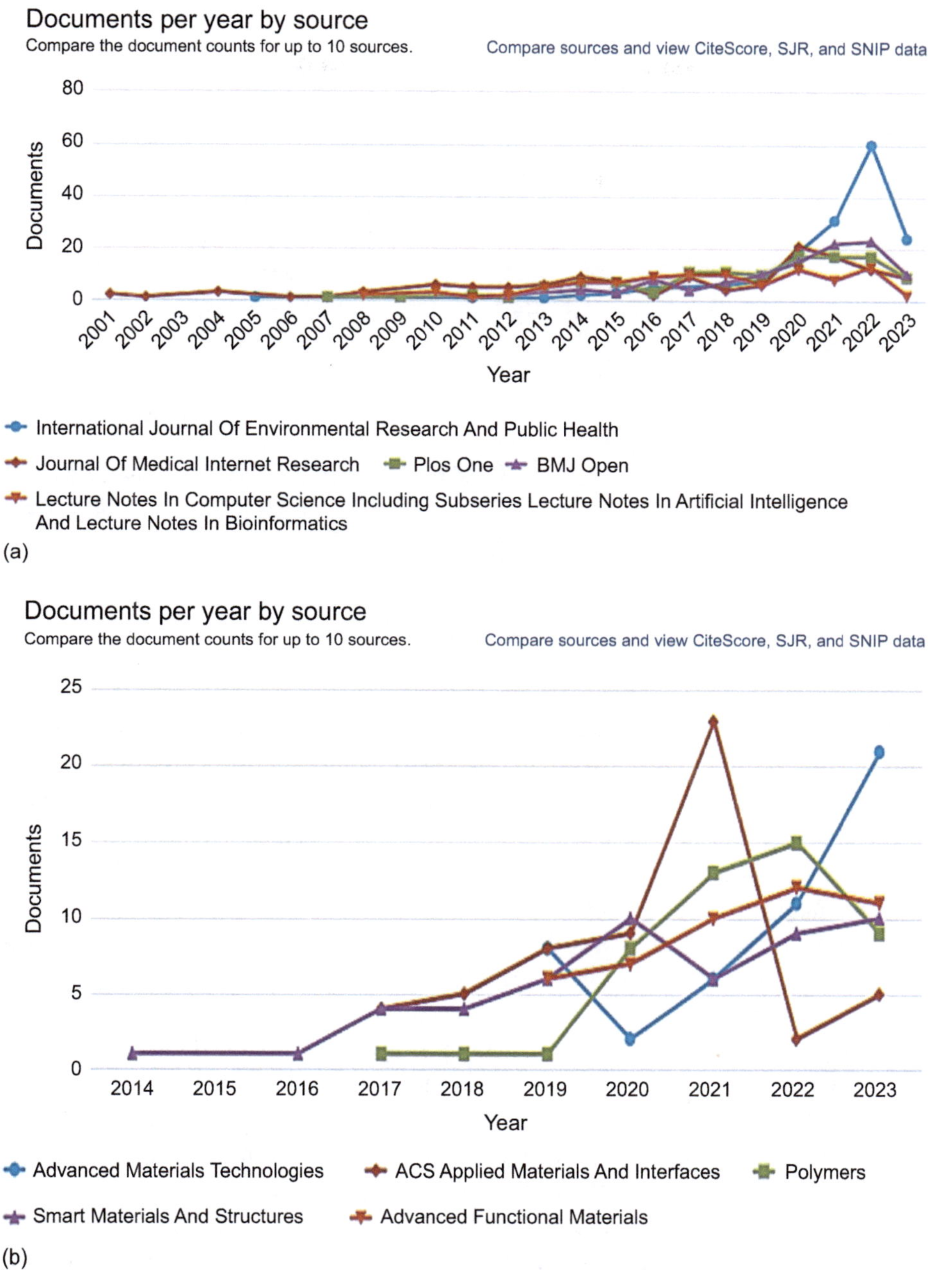

Fig. 4.2: Distribution of published documents by various sources for the past 10 years on (a) advanced applications of smart thermoplastic composites and (b) condition monitoring research areas based on applications of thermoplastic composites.

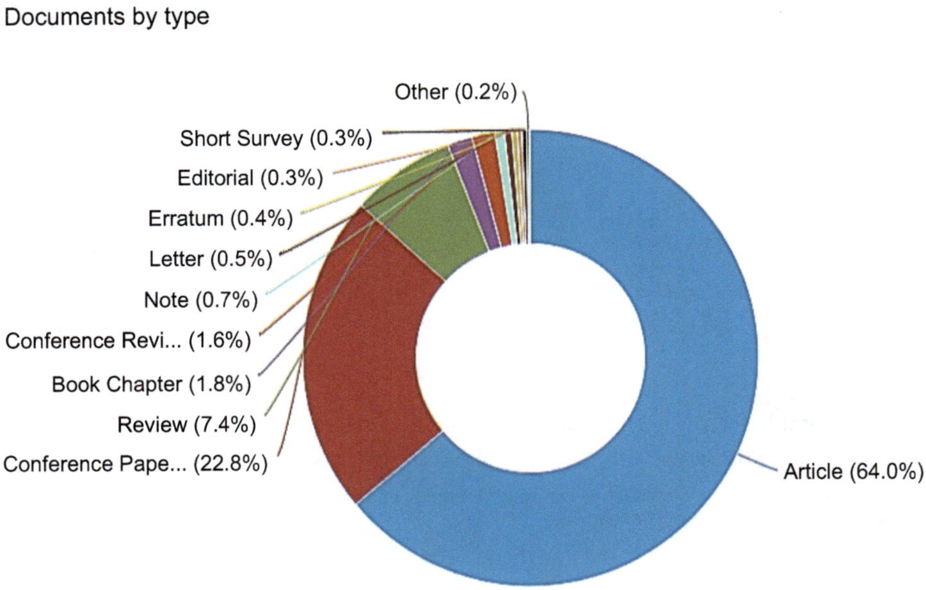

Fig. 4.3: Percentage distribution of various types of documents published on polymer matrix-based composites.

of flexible strain sensors, wearable smart sensors, and 3D-printable conformal devices (Gong et al. 2007, Zhou et al. 2020, Liu et al. 2018). In line with such research activities performed on smart polymer composites for various engineering and biomedical subject areas, Fig. 4.4 shows the pie chart distribution of documents reported on the thermoplastic composite matrix by various subject areas for different consumer applications related to condition monitoring of structural and nonstructural interests.

The exploration of biomedical applications of smart polymer composites for the fabrication of partially absorbable scaffolds, and dental and prosthetic implants reportedly boosted the medicine-based research area of smart thermoplastics (Safian et al. 2021). The investigations performed on two-way programmable repair systems and drug delivery mechanisms have been reported by the fabrication of thermoplastic-based nanocomposites (Bleay et al. 2001, Shalla and Bhat 2021). Smart fabric sensors of polymer matrix like thermoplastic polyurethane also have been explored for human and veterinary patients with acceptable health monitoring capabilities (Aziz and Chang 2018, Singh et al. 2023). Besides the investigations of self-healing and self-repairing features in smart thermoplastic composites for smart drugs and fabrics, the structural health monitoring properties of polymer composites also have been explored by developing the piezoelectric and piezoresistive sensors (Nauman 2021, Balaji and Sasikumar 2016, Leng and Asundi 2003). The experimental investigations performed on polyvinylidene fluoride (PVDF)–SiC-based sensors, 3D-printable multi-material PVDF-based composites, and strain sensors prepared from fused deposition

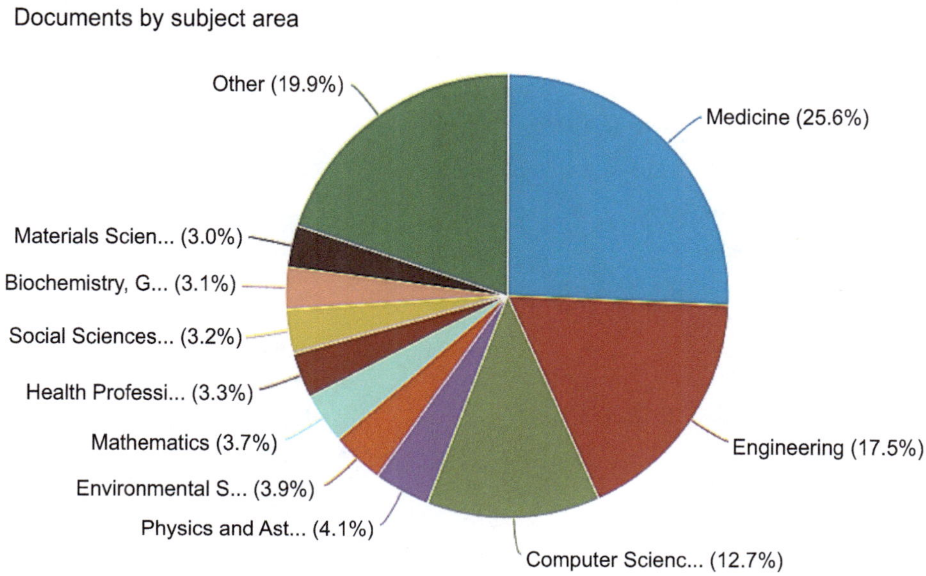

Fig. 4.4: Pie chart distribution of documents reported on the thermoplastic composite matrix by various subject areas.

modeling indicated that the customizable 3D-printed solutions are very useful for structural and nonstructural online health monitoring (OHM) applications (Kumar et al. 2023d, Kumar et al. 2023e, Saleh et al. 2019). Similar to such research activities performed on smart polymer nanocomposites for structural database management and condition monitoring areas, Fig. 4.5 shows the percentage distribution of document types reported on 3D/4D printing-based polymer nanocomposites matrix for condition health monitoring applications.

The role of smart thermoplastic composites has been observed in a variety of biomedical and engineering applications. Electrically conducting graphene, carbon tubes, Cu-doped ZnO, SiC, and so on like reinforcements have been utilized by various investigators to prepare acrylonitrile butadiene styrene, PVDF, and other polymer-based composites for sensing and condition monitoring of various functional systems (Luo and Liu 2014, Kumar et al. 2022b). Some researchers have reported the 4D printing capabilities in polymer blends of polyvinyl chloride and polypropylene for robotics and automation (Ranjan et al. 2023), whereas meta-structure 3D-printed composite-based solutions have also been outlined for structural engineering (Kumar et al. 2022c). As per the reported studies on 3D/4D printing applications of smart thermoplastic composites, Fig. 4.6 shows the distribution of various research documents reported on 4D printing of smart materials by different subject areas for condition monitoring studies as a pie chart.

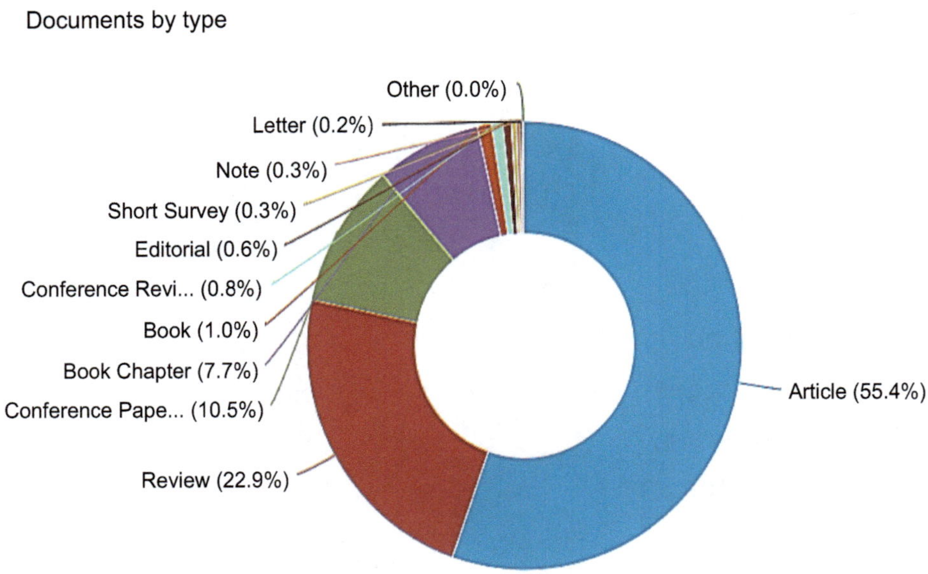

Fig. 4.5: Percentage distribution of document types reported on 3D/4D-printing-based polymer nanocomposite matrix.

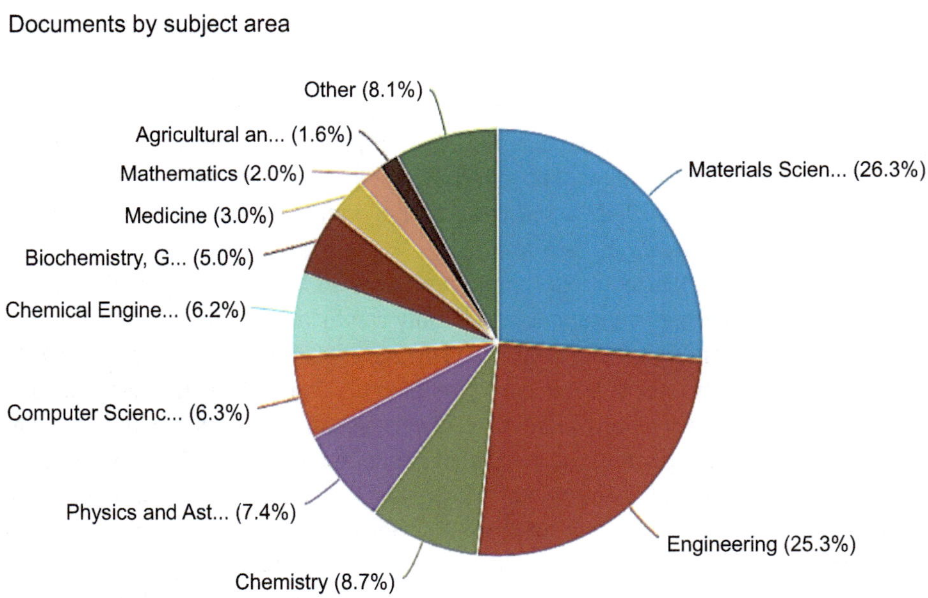

Fig. 4.6: Distribution of documents reported on 4D printing of smart thermoplastic composite matrix by various subject areas in terms of pie chart.

Based on a review of the literature survey, it may be stated that significant work has been reported on different material processing methods for the preparation of smart thermoplastic composites for 3D/4D printing, melt blending, sensor fabrication, and so on, but 4D printing of such smart composites is less reported for OHM applications. The fabrication of standard 4D printing methods for OHM of various systems by using smart thermoplastic composites may not only reduce the operational cost but it will be useful to resolve the environmental issues related to the management of plastic solid waste at a global platform.

4.2 Gap in literature and problem formulation

The review of the literature reveals that a significant number of experimentation-based studies have been explored on novel consumer applications of smart thermoplastic composites for pressure sensing, aerospace, welding of dissimilar polymers, structural monitoring, 3D printing, and so on in the past two decades. However, little work has been performed on 4D printing of smart thermoplastic composites for sensing features based on OHM applications. The research gap analysis, derived from the Web of Science database, was performed to investigate the novel work reported on 4D printing applications of smart thermoplastic composites for OHM. The 2,469 documents obtained (from the Web of Science database) outlined 859 research keywords related to the fabrication of smart polymer-based composites for novel industrial applications like aerospace, sensors, and condition monitoring. The most relevant 81 keywords (out of 284 total reported keywords) given in Tab. 4.1 were processed in the VOSviewer software package to obtain various research areas in which applications of smart composites are highlighted.

On behalf of Tab. 4.1, Fig. 4.7 shows the web of keywords that highlighted the broad area of research studies reported on the preparation of polymer composites for sensing and condition monitoring applications. The web of keywords connected all the explored research aspects by researchers around the globe for different properties of composites desirable for structural and nonstructural engineering applications.

Based on Fig. 4.7, it was observed that polymer composites have been explored extensively for 3D printing-based sensor fabrication as shown in Fig. 4.8a. Similarly, Fig. 4.8b indicated that the properties of polymer composites have been highly investigated for structural condition monitoring (which is also in line with the reported literature). On the other hand, Fig. 4.8c,d outlined the research gap in the previous studies that sufficient work is reported on composite fabrication processes and structural strain monitoring, but lesser work has been reported on testing the 4D printing properties of thermoplastic composites for OHM applications.

Many studies reported recently in the past indicated the 4D printing capabilities in smart thermoplastic composites (Kumar et al. 2023a). 4D properties of PVDF–$CaCO_3$

Tab. 4.1: List of reported technical terms for preparation and applications of smart thermoplastic composites.

S. no.	Terms	Occurrences	Relevance scores
1	Actuator	61	0.9861
2	Advanced composite material	10	0.6781
3	AE signal	15	0.5744
4	Aerospace	27	0.3267
5	Aerospace composite structure	11	1.6802
6	Aerospace industry	13	0.777
7	Aerospace structure	22	0.947
8	Aircraft	73	0.5291
9	Aircraft structure	44	0.7083
10	Algorithm	118	0.7645
11	Amplitude	27	0.5131
12	Art	18	0.3608
13	Artificial intelligence	10	0.8193
14	Attention	26	0.6096
15	Attenuation	15	0.4142
16	Author	23	0.6626
17	Baseline	11	2.4989
18	Beam	33	0.8436
19	Blade	28	0.6402
20	Bridge	51	0.5679
21	BVID tags	10	1.2733
22	Carbon	20	0.596
23	Carbon nanotube	60	2.3533
24	Case study	14	1.3566
25	Cement	19	2.9183
26	CFRP	43	0.2841
27	Civil engineering	11	1.6309
28	Civil infrastructure	13	0.9106
29	CNT	33	2.6683
30	Carbon fibers	15	3.262
31	Complex structure	10	0.5993
32	Complexity	26	0.7094
33	Composite beam	19	1.0959
34	Composite panel	45	0.6112
35	Composite part	16	0.9424
36	Composite patch	16	1.4032
37	Composite plate	72	1.095
38	Composite repair	19	1.4774
39	Concrete	13	2.926
40	Concrete structure	18	1.4676
41	Confidence	11	1.0574
42	Corrosion	18	0.4974
43	Crack growth	15	0.4833
44	Cure	18	0.8649
45	Curing	15	0.9705

Tab. 4.1 (continued)

S. no.	Terms	Occurrences	Relevance scores
46	Cycle	59	0.3878
47	Cyclic loading	14	1.0437
48	Damage assessment	13	0.7482
49	Damage evolution	14	1.0505
50	Damage identification	21	1.7568
51	Damage index	37	1.7193
52	Damage location	17	1.0368
53	Damage state	13	0.6713
54	Damage type	11	1.1807
55	Damaged structure	11	3.0627
56	De-bonding	25	0.4119
57	Deflection	14	1.5089
58	Deformation	45	1.3368
59	Degrees Celsius	24	0.6748
60	Delamination damage	12	0.6808
61	Difficulty	18	1.0271
62	Disbond	28	0.6736
63	Durability	37	0.4033
64	Dynamic response	20	0.9063
65	Edge	17	0.5271
66	Efficiency	28	0.5471
67	Elastic wave	12	0.8411
68	Electrical property	14	2.0915
69	Electrical resistance	30	1.8715
70	Electrode	37	0.9554
71	Electromagnetic interference	10	2.1853
72	Electrospinning-based composites	14	0.5535
73	Environmental condition	17	1.0984
74	Estimation	28	1.4028
75	Evolution	24	0.3977
76	Example	25	0.6641
77	Excitation	43	1.0346
78	Experimental data	16	0.3732
79	Experimental investigation	18	0.4546
80	Extent	24	0.6965
81	Fabrication	38	1.5759

composite and piezoelectric PVDF–graphene–Mn-doped ZnO (PVDF/G/MZ) composites based on 3D-printable solution have been reported for customized conservation of grade III heritage structures. The 4D printing properties of such composites based on their dielectric features may be further explored for OHM of repaired nonstructural cracks of heritage buildings. The study reported in this work is an extension to the previously reported studies on 4D printing of PVDF/G/MZ and PVDF–$CaCO_3$ composites, with a focus on investigating the dielectric sensing properties of such smart composites for OHM of repaired irregularities of heritage building. The 4D-printed

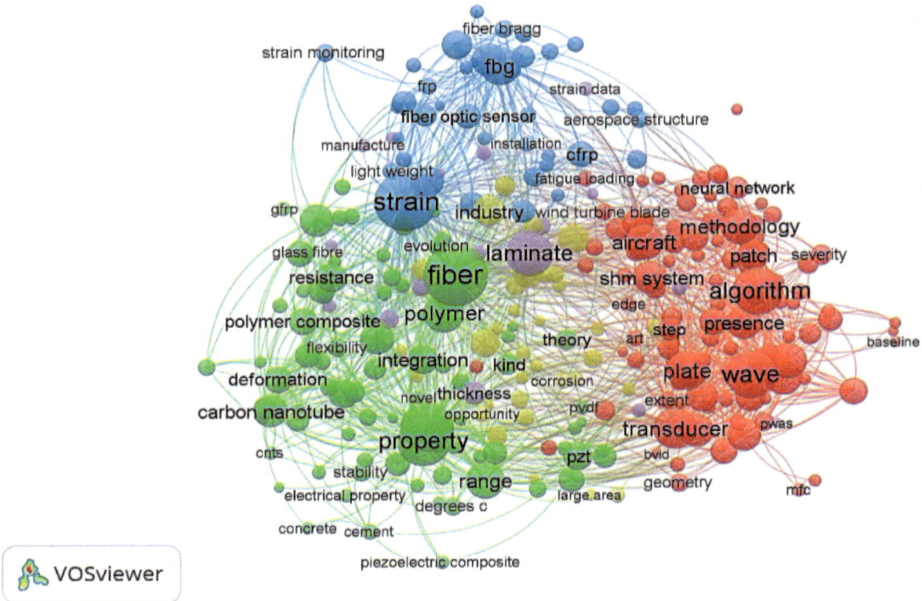

Fig. 4.7: Web of technical words highlighting the wide range of reported work on smart thermoplastic composites.

OHM solution of thermoplastic composite may be used effectively for more accurate preservation and conservation of heritage structures and cultural heritage for future generations.

4.3 Experimentation

The various composite proportions of PVDF/G/MZ and PVDF–$CaCO_3$ were prepared (by mechanical blending process) by varying the weight percentage of reinforcements. Based on the reported literature, the weight proportion of graphene was increased from 0% to 9% and Mn-doped ZnO (from 0% to 4.5%) gradually to obtain nine sets of compositions. On the other hand, the $CaCO_3$ weight proportion was increased from 0% to 9% to obtain different PVDF–$CaCO_3$ compositions/proportions (Kumar et al. 2023e). The compositions so prepared were processed to obtain the feedstock filament by the screw extrusion process, and then the 3D-printable filament samples were tested for mechanical, thermal, rheological, and dielectric (i.e., loss tangent (tan δ) and dielectric constant ($\mathcal{E}_r$)) properties for stimulus-responsive sensor fabrication. The best acceptable composition that possesses good dielectric properties for OHM-based sensing capabilities was 3D printed to obtain the final product for condition monitoring of repaired nonstructural irregularities of heritage structures. The

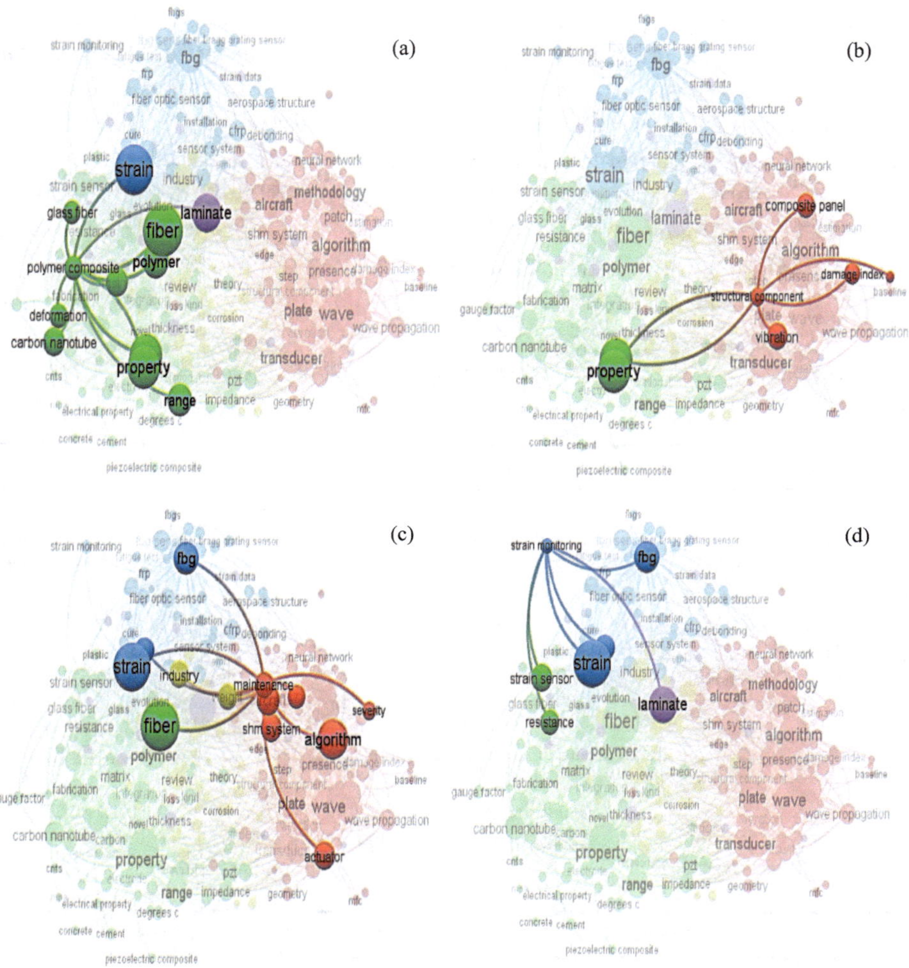

Fig. 4.8: Gap in literature outlining the need for 4D-printed smart thermoplastic composites for online health monitoring (a–d).

multifactor optimization of the feedstock filament fabrication process was performed to ascertain the acceptable material processing parameters for 3D/4D printing of the PVDF-based smart thermoplastic composite matrix.

4.4 Results and discussion

The different compositions prepared for each of the PVDF/G/MZ and PVDF-$CaCO_3$ composites were tested for flowability property-based rheological characteristics in terms of melt flow index (MFI, as per ASTM D1238). Further, the compositions were processed at three different levels of extrusion temperatures ranging from 200 to 220 °C in a twin screw extruder unit (Make: Thermo Fisher miniCTW Haake screw extruder) at three levels of screw speeds (i.e., 100, 110, and 120 rpm) for the preparation of test specimens for mechanical and thermal testing. The dielectric test specimens of each composite proportion were prepared by 3D printing of composites at 220 °C nozzle temperature, 70 °C print platform temperature, and 100% infill density. Table 4.2 shows the results obtained for MFI, Young's modulus (E), heat capacity, ε_r, and tan δ of PVDF/G/MZ composites, whereas Tab. 4.3 shows the results obtained for MFI, E, heat capacity, ε_r, and tan δ of PVDF–$CaCO_3$ composites.

Tab. 4.2: Results obtained for PVDF/G/MZ compositions: rheological, thermal, mechanical, and dielectric property observations.

Composition	Temperature (°C)	rpm	MFI (g/10 min)	HC (J/g)	E (MPa)	ε_r	tan δ
PVDF–1 Gr–0.5 MnZnO	200	100	4.59	38.27	29.4	1.88	0.0055
PVDF–2 Gr–1 MnZnO	210	110	3.78	42.36	31.5	2.15	0.0052
PVDF–3 Gr–1.5 MnZnO	220	120	3.11	48.71	33.2	2.45	0.0051
PVDF–4 Gr–2 MnZnO	200	110	2.66	51.68	35.6	2.83	0.0047
PVDF–5 Gr–2.5 MnZnO	210	120	2.54	54.33	36.2	3.47	0.0046
PVDF–6 Gr–3 MnZnO	220	100	2.41	65.68	38.3	3.73	0.0045
PVDF–7 Gr–3.5 MnZnO	200	120	2.19	66.37	34.1	3.61	0.0034
PVDF–8 Gr–4 MnZnO	210	100	2.01	68.26	33.7	3.37	0.0039
PVDF–9 Gr–4.5 MnZnO	220	110	1.98	69.42	32.8	3.11	0.0040

Tab. 4.3: Observations for rheological, thermal, mechanical, and dielectric properties of PVDF–$CaCO_3$ compositions.

Composition	Temperature (°C)	rpm	MFI (g/10 min)	HC (J/g)	E (MPa)	ε_r	tan δ
PVDF–1% $CaCO_3$	200	100	6.15	41.55	25.4	1.92	0.0054
PVDF–2% $CaCO_3$	210	110	5.36	44.32	29.6	2.19	0.0055
PVDF–3% $CaCO_3$	220	120	4.12	48.09	34.5	2.51	0.0052
PVDF–4% $CaCO_3$	200	110	3.47	52.19	35.9	2.78	0.0048
PVDF–5% $CaCO_3$	210	120	2.76	56.76	37.1	3.41	0.0049
PVDF–6% $CaCO_3$	220	100	2.43	66.57	38.9	3.68	0.0046
PVDF–7% $CaCO_3$	200	120	2.15	67.21	33.2	3.62	0.0033
PVDF–8% $CaCO_3$	210	100	2.02	68.25	33.4	3.28	0.0037
PVDF–9% $CaCO_3$	220	110	1.99	69.04	32.5	3.14	0.0041

The results obtained in Tabs. 4.2 and 4.3 for MFI, heat capacity, E, ε_r, and tan δ of prepared composites outlined that PVDF–6% graphene–3% MnZnO and PVDF–6% $CaCO_3$ possess acceptable flowability characteristics for 3D/4D-printing applications. Based on the acceptable rheological, thermal, mechanical, and dielectric properties, the optimum composite processing parameters were ascertained for the in-house fab-

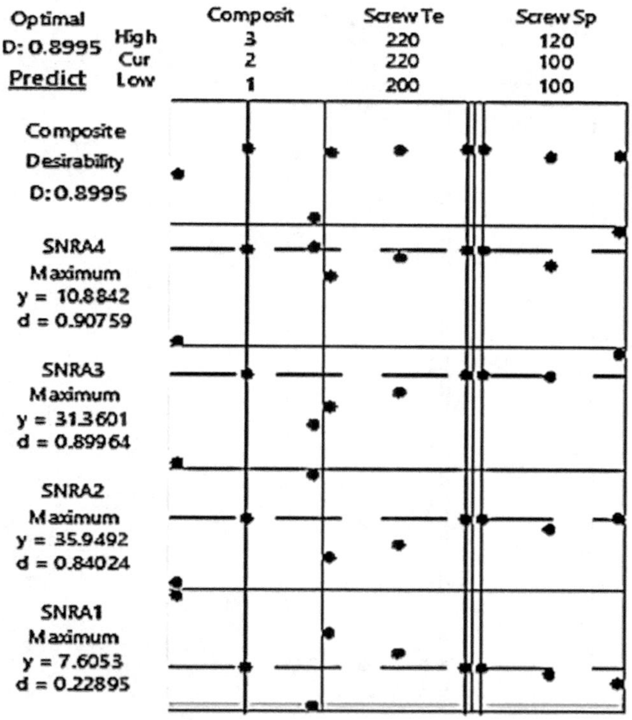

Fig. 4.9: Multifactor optimization observations for PVDF–graphene–MnZnO composite.

rication of the feedstock filaments by screw extrusion process for fused filament fabrication (FFF). The multifactor optimization of composite filament process parameters was performed for both compositions to obtain suitable parameters in context to MFI, heat capacity, E, ε_r, and tan δ. Figure 4.9 shows the multifactor optimization responses of PVDF/G/MZ and Fig. 4.10 shows the multifactor optimization responses obtained for PVDF–$CaCO_3$ composites.

The results obtained for optimization of multifactorial outcomes indicated the composite desirability (D) = 0.8995 for PVDF–graphene–MnZnO case and D = 0.9695 for PVDF–$CaCO_3$, outlining that 220 °C extrusion temperature and 100 rpm extrusion speed are the optimum processing conditions for the fabrication of PVDF-based smart thermoplastic composite for 3D/4D printing of smart sensors for OHM of repaired cracks of heritage structures.

The PVDF matrix-based smart thermoplastic composites were processed at optimized settings for in-house fabrication of the feedstock filament for FFF. The electrically conducting (C) graphene and Mn-doped ZnO-based composite and one-way programmable nonconducting (NC)-type PVDF–$CaCO_3$ composite-based 3D-printed substrate was obtained at 220 °C nozzle temperature, 70 °C bed temperature, and 100% infill density. The 3D printing was performed in a manner to deposit the NC composite layer in be-

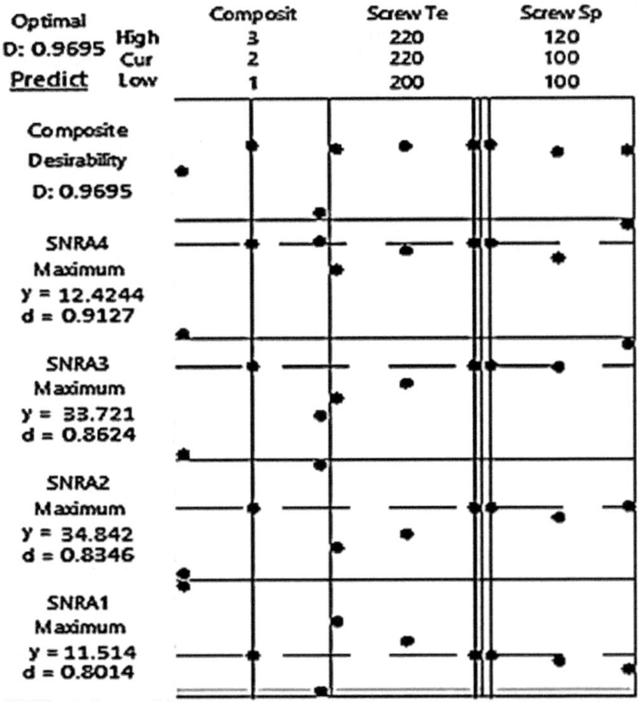

Fig. 4.10: Multifactor optimization observations for PVDF–$CaCO_3$ composite.

tween the C layers of the substrate so that the printed structure responds like a sensor as per the reported literature (Singh et al. 2023). Figure 4.11a shows the cross-sectional view of a 3D-printed sensor with an NC layer sandwiched in the C layer by FFF to obtain the 3D-printed sensor. Figure 4.11b shows the final 3D-printed test specimen to investigate the sensing capabilities for health monitoring of repaired cracks. The insertion loss-based S_{21} parameters of the 3D-printed specimen were investigated by using a vector network analyzer (VNA). The results obtained for VNA-based testing of 3D-printed sensors highlighted that the dielectric properties of NC-type composite significantly contributed to obtaining the radio frequency signals at 2.45 GHz. The ring resonator test also confirmed the dielectric properties of 3D-printed composite structures for 4D-printing-based sensor fabrication. The scanning electron microscopy-based morphological analysis of the substrate surface at 100× magnification (Fig. 4.11c) outlined the uniform distribution of reinforcement to obtain desired sensing properties for online condition monitoring. The surface roughness (Ra) profile of the 3D-printed substrate (Fig. 4.11d) indicated the acceptable Ra = 9.5 μm of the fabricated sensor for the proposed OHM applications in heritage structures.

The study outcomes of the experimental investigation for sensing properties of PVDF matrix-based smart thermoplastic composites outlined that the 4D properties stimulated in the 3D-printed sensor under the effect of the applied dielectric medium

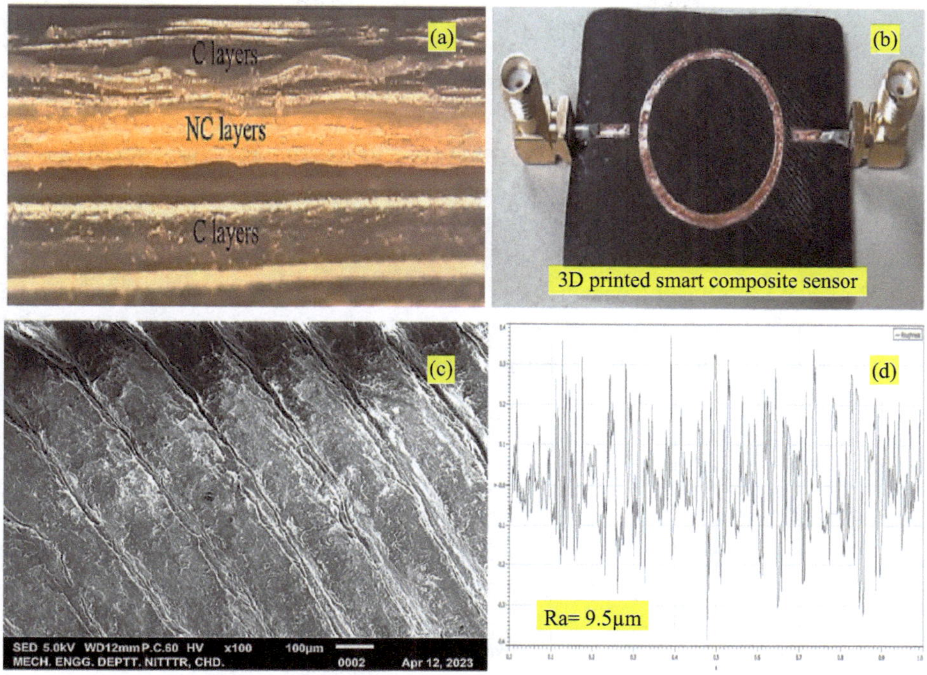

Fig. 4.11: Cross-sectional view of a 3D-printed sensor with an NC layer sandwiched in C layer (a), 3D-printed sensor prototype for S_{21} parameter testing (b), SEM micrograph of the substrate (c), and Ra profile of the substrate (d).

may be applied for a variety of OHM applications like the condition monitoring of repaired structural and nonstructural irregularities like cracks, deterioration, and weathering of heritage buildings. Similarly, the same health monitoring properties of smart thermoplastic composites may be used for biomedical as well as engineering applications.

4.5 Summary

The studies reported on smart thermoplastic composites in the past two decades have outlined very novel industrial applications of such composites for pressure sensing, aerospace, welding of dissimilar polymers, structural monitoring, 3D printing, and so on. But hitherto very little work has been reported on 4D printing of smart thermoplastic composites for sensing features based on OHM applications. This work outlined the investigations for dielectric properties of PVDF-based smart thermoplastic composites for OHM applications in structural and nonstructural engineering. The results of the study show that dielectric properties of the compositions (governed by rheological properties) have a significant effect on mechanical, thermal, morphological, and shape memory properties for OHM of repaired heritage structures. The results obtained for multifactor optimization indicated the composite desirability of 0.8995 and 0.9695 for PVDF–graphene–MnZnO and PVDF–$CaCO_3$, respectively. These also outlined that 220 °C extrusion temperature and 100 rpm extrusion speed are the optimum processing conditions for the fabrication of PVDF-based smart thermoplastic composite for 3D/4D printing of smart sensors. This may be useful for OHM of harmful structural and nonstructural damages faced by heritage structures. It may be stated that the health monitoring properties of smart thermoplastic composites may also be used for biomedical (human and veterinary patients) as well as manufacturing systems, energy storage devices, power generators, and so on based on different engineering setups.

References

Elhajjar R, La Saponara V, Muliana A (Eds.). Smart Composites: Mechanics and Design. 2013, CRC Press, ISBN:9781439895917, 1439895910.

Varga Z, Filipcsei G, Zrínyi M. Smart composites with controlled anisotropy. Polymer. 2005;46(18):7779–87.

Asanuma H. The development of metal-based smart composites. JOM. 2000;52:21–25.

Perez RA, Won JE, Knowles JC, Kim HW. Naturally and synthetic smart composite biomaterials for tissue regeneration. Advanced Drug Delivery Reviews. 2013;65(4):471–96.

Kumar V, Singh R, Ahuja IS. PVDF composite as the 4D solution for heritage structures. Journal of the Institution of Engineers (India): Series C. 2023a;1–14.

Dong B, Li Z. Cement-based piezoelectric ceramic smart composites. Composites Science and Technology. 2005;65(9):1363–71.

Islam MH, Afroj S, Uddin MA, Andreeva DV, Novoselov KS, Karim N. Graphene and CNT-based smart fiber-reinforced composites: A review. Advanced Functional Materials. 2022;32(40):2205723.

Varga Z, Filipcsei G, Szilágyi A, Zrínyi M. Electric and magnetic field-structured smart composites. In: Macromolecular Symposia. WILEY-VCH Verlag, Weinheim, 2005, Vol. 227, No. 1, 123–34.

Kumar V, Singh R, Ahuja IS. In-house development of smart materials for 4D printing. In: 4D Imaging to 4D Printing. CRC Press, 2022a, 85–102.

Lau KT. Fiber-optic sensors and smart composites for concrete applications. Magazine of Concrete Research. 2003;55(1):19–34.

Tauchert TR, Ashida F, Noda N, Adali S, Verijenko V. Developments in thermos piezoelasticity with relevance to smart composite structures. Composite Structures. 2000;48(1–3):31–38.

Pucci A, Bizzarri R, Ruggeri G. Polymer composites with smart optical properties. Soft Matter. 2011;7 (8):3689–700.

Kumar V, Singh R, Kumar R, Ranjan N, Hussain M. ABS-PLA-Al composite for digital twinning of heritage buildings. Materials Letters. 2023b;346:134536.

Umar K, Yaqoob AA, Ibrahim MNM, Parveen T, Safian MTU. Environmental applications of smart polymer composites. In: Smart Polymer Nanocomposites. Woodhead Publishing, 2021, 295–312.

Ikram H, Al Rashid A, Koç M. Additive manufacturing of smart polymeric composites: Literature review and future perspectives. Polymer Composites. 2022;43(9):6355–80.

Kumar V, Singh R, Ahuja IS. 3D printed thermoplastic composite-based Innovative solutions for heritage structures: A review on technology to application. Journal of the Institution of Engineers (India): Series C. 2023c; 1–22.
Gong B, Sanford AR, Ferguson JS, Filipcsei G, Csetneki I, Szilágyi A, Zrínyi M. Magnetic field-responsive smart polymer composites. In: Oligomers-polymer Composites-molecular Imprinting. 2007, 137–89.
Zhou K, Dai K, Liu C, Shen C. Flexible conductive polymer composites for smart wearable strain sensors. SmartMatter. 2020;1(1):e1010.
Liu H, Li Q, Zhang S, Yin R, Liu X, He Y, Guo Z. Electrically conductive polymer composites for smart flexible strain sensors: A critical review. Journal of Materials Chemistry C. 2018;6(45):12121–41.
Safian MTU, Umar K, Parveen T, Yaqoob AA, Ibrahim MNM. Biomedical applications of smart polymer composites. In: Smart Polymer Nanocomposites. Woodhead Publishing, 2021, 183–204.
Bleay SM, Loader CB, Hawyes VJ, Humberstone L, Curtis PT. A smart repair system for polymer matrix composites. Composites Part A: Applied Science and Manufacturing. 2001;32(12):1767–76.
Shalla A, Bhat M. Smart polymer composites in drug delivery. In: Smart Polymer Nanocomposites. Woodhead Publishing, 2021, 261–94.
Aziz S, Chang SH. Smart-fabric sensor composed of single-walled carbon nanotubes containing binary polymer composites for health monitoring. Composites Science and Technology. 2018;163:1–9.
Singh R, Dhaliwal BS, Pattnaik SS (Eds.). 3D printing of sensors, actuators, and antennas for low-cost product manufacturing, 2023.
Nauman S. Piezoresistive sensing approaches for structural health monitoring of polymer composites – A review. Engineering. 2021;2(2):197–226.
Balaji R, Sasikumar M. Structural health monitoring (SHM) system for polymer composites: A review. Indian Journal of Science and Technology. 2016;9(41):1–12.
Leng J, Asundi A. Structural health monitoring of smart composite materials by using EFPI and FBG sensors. Sensors and Actuators A: Physical. 2003;103(3):330–40.
Kumar V, Singh R. Investigation on 3D printed PVDF–SiC composites for phagocytosis sensor application. National Academy Science Letters. 2023d;1–5.
Kumar V, Singh R, Ahuja IS. On multi-layered PVDF composite-based 3D-printed solution for repair of non-structural cracks in heritage buildings. National Academy Science Letters. 2023e;1–5.
Saleh MA, Kempers R, Melenka GW. 3D printed continuous wire polymer composites strain sensors for structural health monitoring. Smart Materials and Structures. 2019;28(10):105041.
Luo S, Liu T. Graphite nanoplatelet enabled embeddable fiber sensor for in situ curing monitoring and structural health monitoring of polymeric composites. ACS Applied Materials & Interfaces. 2014;6 (12):9314–20.
Kumar V, Singh R, Ahuja IPS. On 4D capabilities of chemical assisted mechanical blended ABS-nano graphene composite matrix. Materials Today: Proceedings. 2022b;48:952–57.
Ranjan N, Kumar R, Singh R, Kumar V. On PVC-PP composite matrix for 4D applications: Flowability, mechanical, thermal, and morphological characterizations. Journal of Thermoplastic Composite Materials. 2023;36(4):1401–21.
Kumar V, Singh R, Ahuja IS. On 3D printed meta-structure-based functional prototype as an innovative solution for repair and online health monitoring of heritage structures. Materials Letters. 2022c;326:132950.

Bharat Kalia*, Rupinder Singh, J. Paulo Davim

5 Development of 3D metal-printed smart dental implants

Abstract: Additive manufacturing, particularly metal 3D printing, has transformed the dental implant business by providing precise and personalized solutions. Among the several processes, direct metal laser sintering (DMLS) stands out as a versatile and exceptionally accurate method for creating dental implants. DMLS technology allows for the precise manufacturing of dental implants, allowing for a perfect fit within the patient's oral anatomy. Furthermore, the ability of DMLS to build complex geometries promotes osseointegration and shortens healing times. Multimaterial dental implant manufacturing is gaining pace in addition to single-material dental implant manufacturing. Combining materials of varying quality in a single implant improves load distribution, stability, and aesthetics. This unique approach boosts patient satisfaction while also increasing the life of dental implants and long-term stability. In the recent past, researchers have utilized the 3D printing process to fabricate dental implants using a single material, but hitherto little has been reported on fabricating dental implants using multiple materials. For developing smart property within an implant, the most compatible materials are shape memory alloys. But the same smartness can also be achieved by using kirigami-based megastructures. Thus, in this chapter, DMLS technique was used for developing multimaterial smart dental implants with metastructure.

***Corresponding author: Bharat Kalia**, Department of Mechanical Engineering, National Institute of Technical Teachers Training and Research, Chandigarh, India, e-mail: bharatkalia98@gmail.com
Rupinder Singh, Department of Mechanical Engineering, National Institute of Technical Teachers Training and Research, Chandigarh, India
J. Paulo Davim, Department of Mechanical Engineering, University of Aveiro, Portugal

https://doi.org/10.1515/9783111081632-005

Graphical abstract:

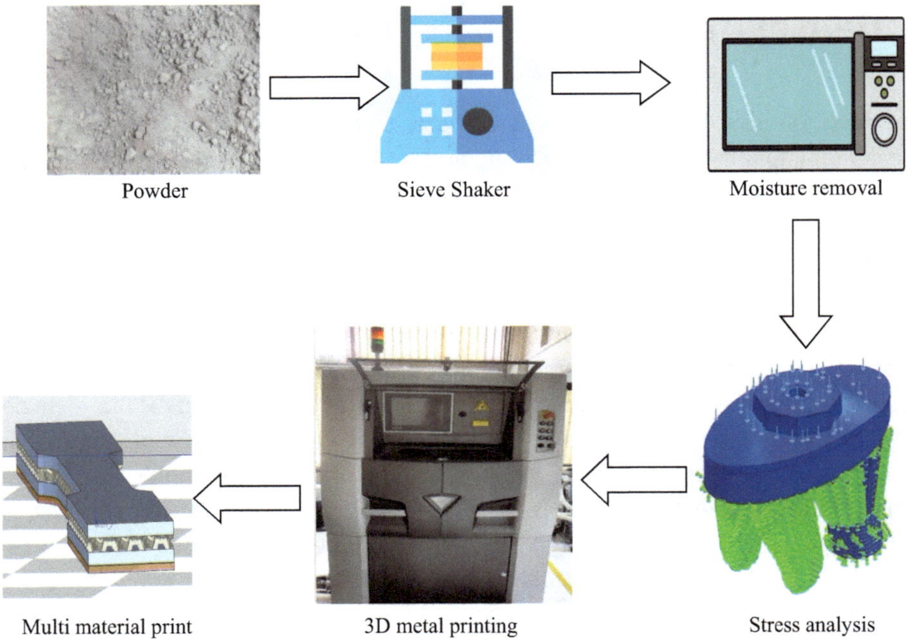

Keywords: Additive manufacturing, direct metal laser sintering, smart dental implants, multimaterial

5.1 Introduction

3D printing, often known as AM, is an advanced manufacturing method (Lukic et al. 2016). It uses standardized materials for manufacturing customized 3D items using predetermined automatic processes based on digital computer-aided design (CAD) models (Vukicevic et al. 2017, Alharbi et al. 2018, Lin et al. 2019). It may convert representations of geometry into actual physical items through the addition of material in layers. Many 3D procedures have grown dramatically in recent years. In the year 1980, Charles Hull made 3D printing technologies publicly available (Holzmann et al. 2017).

3D printing has grown widely in recent years and is employed for quick prototyping in a variety of industries like manufacturing design and engineering. In the rapidly developing field of new materials, machines and technologies will be completely changed by 3D printing (Gross et al. 2014). The development of the technology for 3D printing has been incredibly inventive and versatile. It opens up new alternatives for firms attempting to boost industrial efficiency and encourages many others.

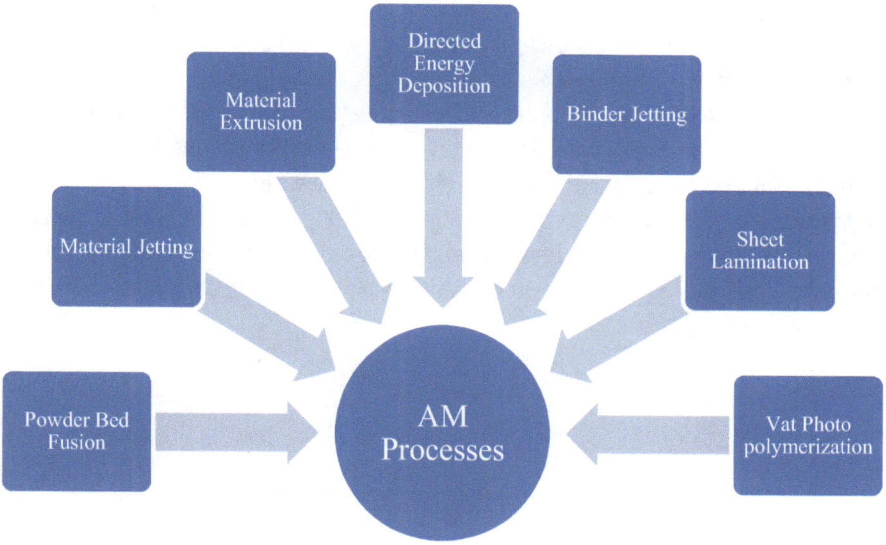

Fig. 5.1: Classification of AM processes.

As per ASTM standards, AM processes are classified into seven categories (Fig. 5.1). There are different process parameters for AM processes; however resolution and fabrication speed are most critical. In the past decade, a lot of work has been reported on the 3D printing of dental implants using a single material, but hitherto little has been reported on fabricating dental implants using multiple materials. For generating smartness within an implant the most suitable materials are shape memory alloys. But the same smartness can also be achieved by using kirigami-based metastructures. In this chapter, the DMLS technique was used for developing a multimaterial smart dental implant with metastructure.

5.2 Metal 3D printing

A CAD file was processed step by step to produce the 3D model in the additive fabrication process, commonly referred to as 3D printing. This method can be used to create highly customized items that would be difficult to create using regular manufacturing methods. Other benefits of AM include its high material utilization, low fixed costs, lack of labor intensiveness, and overall environmental friendliness. Modern AM technology can be employed with a wide variety of materials like metals, polymers, composites, and ceramics. Perhaps the most significant impact among them has been demonstrated by AM of metallic materials across numerous areas such as aviation, automotive, and biomedical (Sames et al. 2016, Macdonald et al. 2016, Smith et al. 2018,

Bose et al. 2018, Harun et al. 2018). The advantages of adopting AM technologies to manufacture metallic materials include the capacity to develop and create compositions with specialized properties using monolithic, multimaterial compositions, as well as the ability to fabricate complex shapes (Zhang and Bandyopadhyay 2018, Bandyopadhyay and Heer 2018, Zhang and Bandyopadhyay 2019). AM methods for metallic materials can be categorized into different groups according to the various kinds of feedstock materials and energy supply indicated in Fig. 5.2. Powder and wire material sources are extensively employed in metal AM technologies. The wire serves as the feedstock, and this equipment can employ electron, laser, or plasma arc beams as their source of energy.

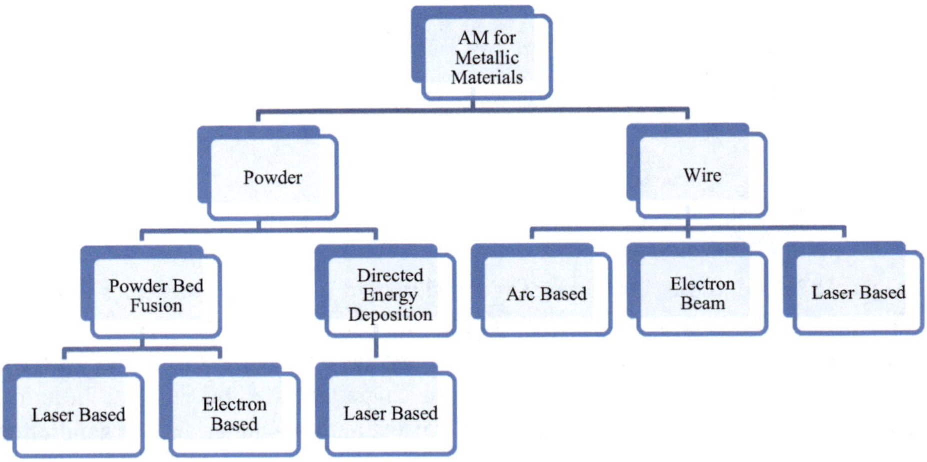

Fig. 5.2: Categorization of different AM techniques.

Powder bed fusion (PBF) selectively melts or sinters a metal powder bed in powder feed systems using a high-energy power source. According to the type of power source employed, a powder feed system can be separated into two major methods. The first is the process of selective laser melting (SLM) making use of a high-intensity laser and the second is electron beam melting (EBM) using an electron beam. A platform is required to support the powder during both procedures. The most extensively used techniques in additive manufacturing (AM) for metals are SLM, direct metal laser sintering (DMLS), and direct energy deposition (Rao and Rao 2021).

5.3 Direct metal laser sintering

The widely used PBF technique for manufacturing solid metal parts employing a layer-by -layer method using metal powders is called DMLS. This technique is part of AM. In this method, individual parts are created layer by layer using metal powders. Manufacturing, aerospace, automotive, naval, biomedical, and other industries use this method significantly. When the University of Texas began exploring the use of Selective Laser Sintering (SLS) technology on metals in 1990, the second PBF technique was developed. Although the initial attempts to use zinc or lead were unsuccessful, they dubbed the technique as DMLS. However, the Fraunhofer Institute of Production Technology reported the initial application of 316L stainless steel powder a success in 1994 (Vollertsen et al. 1994).

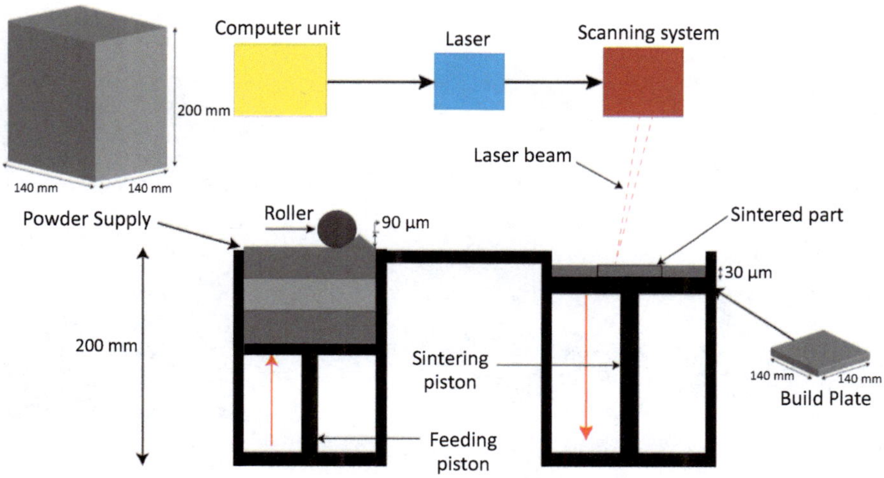

Fig. 5.3: Schematic of DMLS process.

As the metal particles do not entirely melt during this technique, a relatively low laser power is often employed. High temperatures are required for this method because the powder must be heated fast (Sanz and Navas 2013, Zhao et al. 2017, Zhang and Zhang 2017, Keshavarzkermani et al. 2018, Gajera et al. 2019). The DMLS process begins with the application of a powder layer to the substrate, which is subsequently held in place on the construction platform. The laser scans the complete surface. During the process, the laser fuses the powder, causing the powder bed to melt selectively or partially. Following that, the powder bed is crushed to form a dense component. As a result, the consolidation process is semisolid. The platform is subsequently lowered by the dimension of the layer. To get the final product, a fresh covering of powder is placed, and the procedure is repeated. Figure 5.3 depicts the DMLS operation schematic and Fig. 5.4 illustrates the different setups required for printing parts on the DMLS process starting from powder preparation and final print to postprocessing for removing residual stresses.

Porous implants with close-to-bone mechanical properties (stiffness) can be manufactured using DMLS technology. In actuality, a porous implant will have a lower elastic modulus and yield strength than a fully dense component. In orthopedics, stress shielding causing bone resorption presents a significant challenge because it is thought to be an important factor in unwanted results such as aseptic implant loosening (Hollander et al. 2006, Mullen et al. 2009, Stamp et al. 2009, Mangano et al. 2009, Dabrowski et al. 2010, Fukuda et al. 2011). One way to lower stress shielding is to employ porous metallic biomaterials, which have a high elastic modulus when a large number of linked pores are put into them (Traini et al. 2008). Unwanted effects such as stress shielding can thus be avoided by modifying the mechanical properties of porous devices to more nearly mimic the elastic modulus and yield strength of recipient bone.

Fig. 5.4: DMLS process setup.

5.4 Materials used in the DMLS process

5.4.1 Titanium alloys

In powder-based AM, Ti6Al4V, the most prevalent titanium alloy, accounts for over 50% of the world's titanium product market share. It offers exceptional mechanical qualities and corrosion resistance. It is a strong but lightweight alloy that is ideal for aviation engines, gas turbines, and a range of airplane parts (Boyer 1996, Inagaki et al. 2014). While the aerospace industry continues to dominate Ti6Al4V demand, additional usage domains including marine, vehicle, energy, chemical, and biological have gained widespread acceptance of Ti6Al4V due to low density, excellent strength, good corrosion resistance, and biocompatibility. Due to its well-studied biocompatibility, it is also used in the biomedical sector, particularly for limb prostheses and orthopedic implants (Niinomi et al. 2012).

5.4.2 Stainless steel alloys

17-4 PH stainless steel (AISI 630 stainless steel) is a significant engineering alloy (Viswanathan et al. 1988, Hsiao et al. 2002) that has captured the attention of laser PBF researchers. SS17–4PH has exceptional weldability, corrosion resistance, and mechanical qualities. The term precipitation hardening is indicated by the acronym "PH" in 17-4 PH stainless steel. To enhance its mechanical qualities, this alloy goes through a specific heat treatment method known as aging or precipitation hardening. After heat treatment, this material can exhibit remarkable mechanical properties such as high tensile/impact strength, toughness for fractures, and resistance to corrosion at temperatures of use below 300 °C (Schade 2010). Grades like 316L, 304L, and 17-4 stainless steel alloys are extensively employed in medical, automobile, aviation, and tooling industries for various applications like dental goods, hip and knee joints, casting, and extrusion dies.

5.4.3 Nickel-based alloys

Nickel alloys are available in a variety of chemical compositions, microstructures, and mechanical characteristics. Nickel-based superalloys are a type of precipitation-hardenable alloy that behaves well at high temperatures (Kudzal et al. 2017, Zhang et al. 2021). Nickel-based superalloys have been employed in various fields due to remarkable high-temperature characteristics, including aviation, reactors for nuclear energy, thermal power plants, rocket engines, and so on. In general, they are more resistant to corrosion than stainless steel. Nickel alloy 625 has excellent fatigue and thermal fatigue properties, as well as excellent corrosion resistance. Nickel alloy 718 is a precipitation-hardenable Ni–chromium alloy containing a trace of iron, niobium, molybdenum, aluminum, and titanium.

5.4.4 Cobalt-chrome alloys

Cobalt-chrome alloys often referred to as CoCr alloys are a group of metallic alloys known for their excellent properties, making them valuable for various industries and medical applications. Alloys of cobalt–chromium are biocompatible with high chromium content and good tensile, corrosion, and fatigue properties. (Ameer et al. 2004, Gaytan et al. 2011, Xin et al. 2013, Monroy et al. 2013, Monroy et al. 2014). Due to their outstanding mechanical qualities, CoCr alloys are becoming more popular for dental restorations (Fu et al. 2022). These alloys are popularly used in the dentistry field for metal frames, partial dentures, and dental restorations. Recently, they have been employed to make stents and for body joints such as knees and shoulders.

5.5 Bibliographic analysis and research gap

Figure 5.5 shows the basic classification of materials used for AM. The Web of Science database was explored for the keyword 3D metal-printed dental implant. The information from these selected articles was entered into the VOSviewer software. By limiting the number of occurrences to five, 159 terms out of 4,545 met the requirement. Furthermore, a relevance score was derived for these 159 terms, and the top 60% (95) terms were used for the study, as shown in Fig. 5.6.

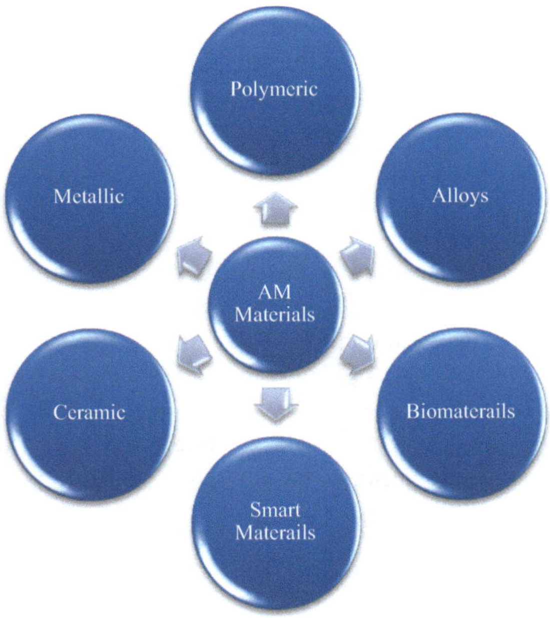

Fig. 5.5: Classification of materials used for AM.

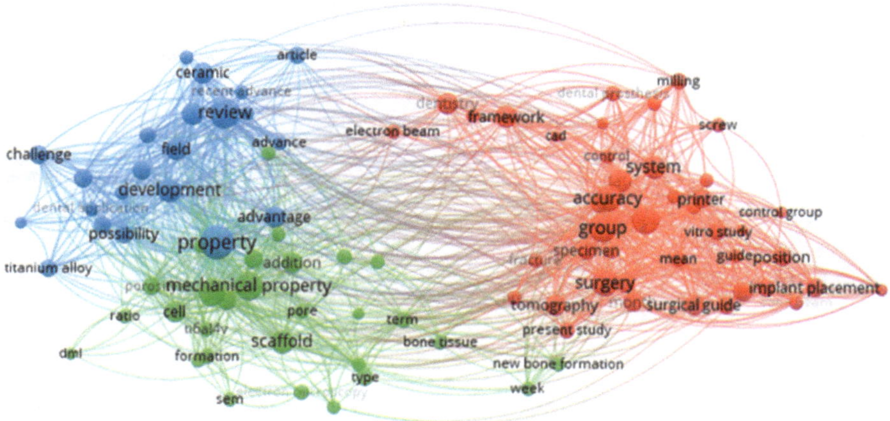

Fig. 5.6: Reported research on 3D metal-printed dental implant using open-source VOSviewer software.

Based on bibliographic analysis the research gaps in the form of network diagrams are shown in Figs. 5.7 and 5.8.

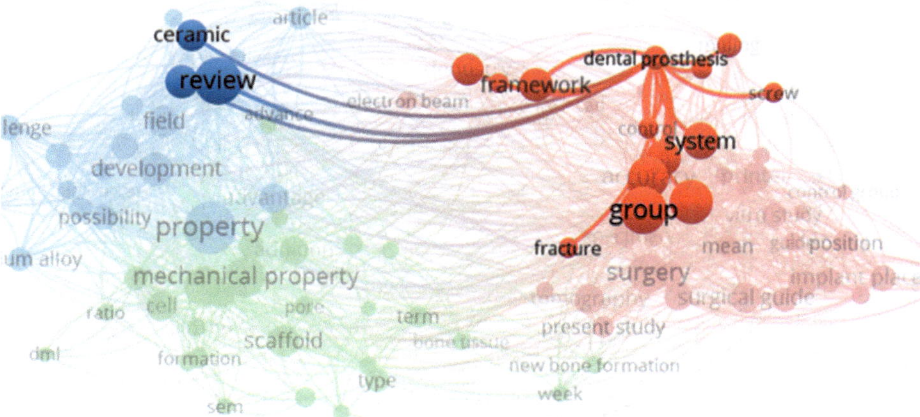

Fig. 5.7: Gaps in previously reported literature related to dental prosthesis and DMLS.

Table 5.1 shows the studies reported on 3D metal-printed dental implants with a minimum of five occurrences.

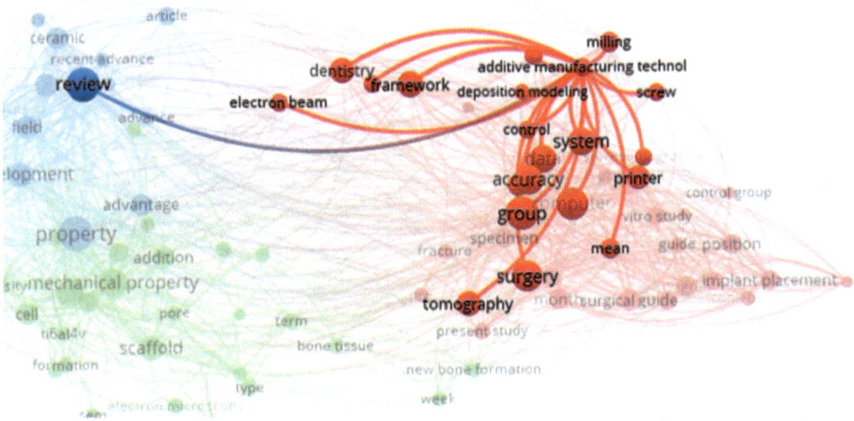

Fig. 5.8: Gaps in previously reported literature related to additive manufacturing and dental implant.

Tab. 5.1: Occurrence and relevance score for the keyword "3D metal-printed dental implant."

S. no.	Term	Occurrences	Relevance score
1	Accuracy	22	0.5949
2	Addition	14	0.4462
3	Additive manufacturing technology	7	1.2275
4	Advance	7	0.6414
5	Advantage	15	0.5351
6	Alloy	23	0.6938
7	Alpha	6	1.2921
8	Article	10	0.659
9	Biocompatibility	11	0.5634
10	Biomedical application	12	1.724
11	Bone regeneration	7	0.4044
12	Bone tissue	6	0.6862
13	Cad	5	0.5014
14	Cell	14	1.0902
15	Ceramic	14	0.8401
16	Challenge	11	2.0444
17	Chemical composition	6	0.8268
18	Comparison	10	1.5339
19	Composite	9	1.0889
20	Computer	25	0.7229
21	Cone beam	7	1.9778
22	Control	6	0.7629
23	Control group	5	1.955
24	Data	17	0.8167
25	Degree	13	1.7079

Tab. 5.1 (continued)

S. no.	Term	Occurrences	Relevance score
26	Dental application	7	1.1225
27	Dental prosthesis	8	0.9753
28	Dentistry	14	0.4085
29	Deposition modeling	5	0.7277
30	Development	18	0.918
31	Difference	12	0.8003
32	Digital workflow	5	1.3068
33	Dml	5	2.5599
34	EBM	6	0.4341
35	Electron beam	7	0.3613
36	Electron microscopy	8	0.9332
37	Example	6	1.1796
38	Field	16	0.8573
39	Finite element analysis	6	0.5056
40	Formation	8	1.3338
41	Fracture	6	0.6192
42	Framework	16	0.3772
43	Group	28	0.5742
44	Guide	7	1.4151
45	Implant placement	12	2.1533
46	Industry	12	1.1164
47	Mean	9	1.2362
48	Mechanical property	24	0.6223
49	Mechanical strength	7	0.9261
50	Metal sleeve	5	2.9828
51	Metallic implant	5	1.8885
52	Microstructure	12	0.8001
53	Milling	9	1.213
54	Month	13	1.0019
55	New bone formation	6	0.6454
56	Overview	7	0.9954
57	Peek	6	0.6134
58	Polymer	16	0.8604
59	Pore	7	0.6753
60	Porosity	10	1.1923
61	porous structure	8	0.4976
62	Position	11	1.8956
63	Possibility	11	0.9721
64	Pre	6	0.8061
65	Present study	9	0.6412
66	Printer	13	1.0978
67	Property	36	0.6385
68	Ratio	6	1.6223
69	Recent advance	6	0.7609
70	Recent year	7	0.7371
71	Resin	7	1.4224
72	Review	29	0.7281

Tab. 5.1 (continued)

S. no.	Term	Occurrences	Relevance score
73	Scaffold	17	0.5649
74	Screw	6	1.5754
75	Selective laser sintering	5	0.4959
76	Sem	6	1.776
77	Significant difference	10	1.9428
78	SLA	6	0.7307
79	Specimen	11	0.7935
80	Stereolithography	6	1.0609
81	Surface roughness	7	0.8412
82	Surgery	23	0.4973
83	Surgical guide	15	1.0204
84	System	18	0.7904
85	Tensile strength	6	0.5928
86	Term	9	0.2797
87	Ti6Al4V	7	0.9843
88	Tissue engineering	10	1.3477
89	Titanium alloy	10	1.8205
90	Tomography	14	0.4113
91	Tooth	10	0.6979
92	Type	10	0.483
93	In vitro study	10	1.1848
94	Week	6	0.6194

5.6 Case study: single material dental implant

5.6.1 Background

Teeth are one of the important organs that carry out a variety of tasks in healthy animals (Coffman et al. 2019). Canines exhibit diphyodont development and heterodont tooth forms (Capik et al. 2000). According to canine literature, 28 primary teeth erupt by the age of 3–6 months, which are taken over by 42 permanent teeth by the age of 6–7 months. Tooth fractures are pretty common due to involvement in regular tasks and fractures in strategic teeth (ST) are relatively prevalent. The animals become ineffective and cease eating when the ST is lost completely or partially. As a result, ST protection is critical for the canine's health.

In the modern era, computer-aided modeling and 3D printing are being used for effectively tackling the complex issues associated with patient-specific orthotic and prosthetic solutions. These advanced techniques involve the use of digital design and fabrication processes to create customized orthoses and prostheses tailored precisely to each patient's unique needs. This innovative approach not only enhances the overall fit and comfort of such devices but also improves their functional performance, ultimately leading to better patient outcomes and quality of life. DMLS is an innovative and nontraditional method within the realm of metal AM. This cutting-edge technique enables the creation of implants with exceptional precision and customization. Unlike conventional implant fabrication methods, which often involve extensive machining or casting processes, DMLS offers a more versatile and precise approach. Metal powder is melted progressively in conjunction with CAD model data to create metallic parts of any desired shape (Sumanth and Srinivasan 2022). Typically, a porous structure is created using DMLS. The elastic modulus of implant material is critical to prolonged implant success since dissimilarity in value may lead to stress shielding (Chang et al. 2019). DMLS allows for the manufacturing of functionally graded implants, which are always preferable because they match the elastic properties of bones and minimize the stress-shielding effect ensuring long-term stability of the implant (Traini et al. 2008).

Animals bite harder than humans; nature has designed some animal teeth with three roots to withstand the additional strain. Veterinary patients face more impediments to accessing dental care than human patients, which commonly results in untreated or extracted teeth due to pain or discomfort. In the multirooted tooth family, the canine is the only single-rooted tooth, while the first lower molar has two roots and the final upper premolar has three roots. The current approach to treating multirooted dental disorders requires time-consuming treatments such as bone grafting surgery on all root surfaces and subsequent drilling for a new implant. As a result, there is an urgent need for a one-of-a-kind and original design for a multirooted dental implant. This work focuses on the creation of a commercially viable 3D-printed functioning prototype (Fig. 5.9). This multirooted dental implant (MRDI) concept is specifically intended for canines and is created utilizing DMLS equipment and metallic alloy powder, 17-4PH SS.

5.6.2 Fractured canine tooth

The right maxillary PM4 tooth of a three-year-old male German Shepherd was selected for multirooted implant design.

5.6.3 3D scanning

The tooth was scanned using a dental lab scanner. Further, the data generated was processed on 3D sprint basic software for 3D modeling. Using the measurement tools embedded in the software, diverse sets of dimensional data were collected to be used

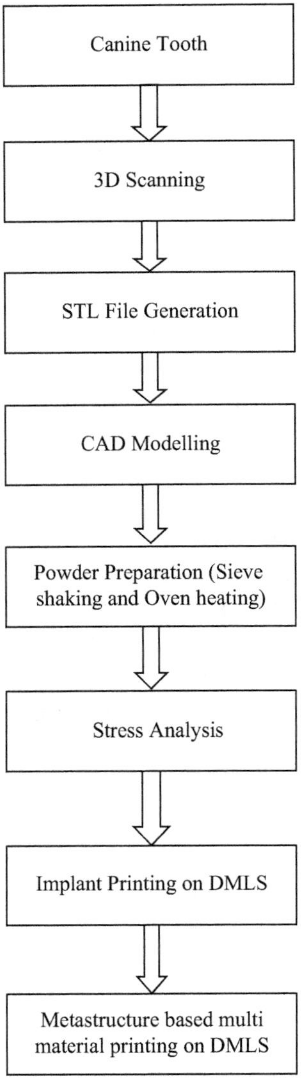

Fig. 5.9: Process flow diagram.

in the design of the implant. The design strategy adhered to the guideline outlined by Coffman et al. (2019), which stipulates that approximately two-thirds of the implant's total length should constitute the root, while the remaining one-third should form the crown. Subsequently, this comprehensive set of dimensional data was employed to create a detailed 3D model, which was then saved in the STL file format. This STL file serves as the foundation for subsequent phases of processing and development in the implant design workflow.

5.6.4 Implant CAD modeling

The primary objective in implant design is to replicate the structure of the natural tooth, such as the PM4 tooth in a canine, which features three distinct roots. During the implant design process, particular attention was dedicated to addressing three critical factors: achieving primary stability, ensuring secondary stability, and mitigating issues related to screw loosening. These factors were prioritized due to their significant association with common causes of implant failure. The MRDI assembly, in this context, consists of a total of eight individual components.

5.6.5 Stress analysis

Using the SOLIDWORKS simulation package, a FEM-based stress analysis was conducted (Fig. 5.10). This analysis involved the assembly of implant components, including the abutment, into a single unit for modeling purposes. To simulate the osseointegrated behavior of the implant, fixed bond state was established at the interface between the implant and surrounding bone tissue.

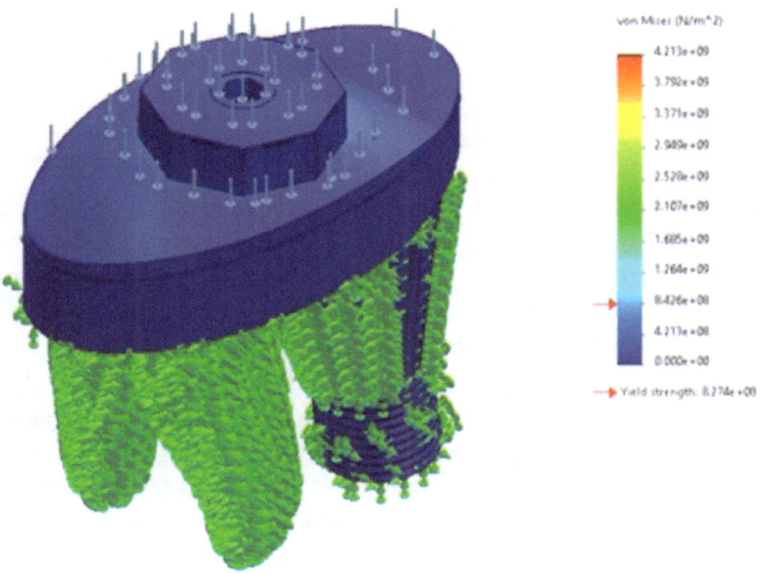

Fig. 5.10: Stress analysis of assembly.

The 3D model was discretized using a mesh composed of solid tetrahedral elements characterized by blended curvature. An axial static load of 800 N was applied to assess the stress distribution within the implant assembly. The Von Mises stress analysis highlighted that the designed implant assembly operates within safe parameters and can withstand an imposed force of 800 N as shown in Fig. 5.10.

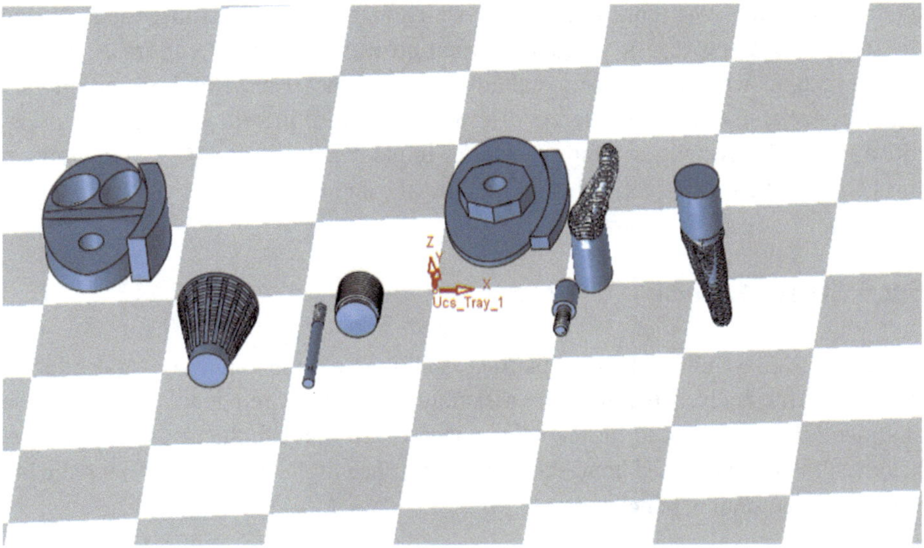

Fig. 5.11: Implant assembly parts on 3DXpert software.

5.6.6 DMLS printing of implant

Following the analysis of individual sub-assemblies, these components were imported into the 3DXpert software by 3D Systems, where they were arranged on a build plate with dimensions of $140 \times 140 \times 100$ mm^3. Subsequently, adjustments were made to determine the orientation, type of support structures, and the height at which these supports would be applied. On this software platform, a virtual printable CAD model was generated, which served as the blueprint for the actual printing process (Fig. 5.11). The physical realization of this design was achieved using a ProX DMP 200 metal printer, specifically configured for dental applications. The material of choice for this process was 17–4 precipitation-hardened stainless steel grade B powder, adhering to standard inbuilt settings and dental configuration established for this printer.

The dental implant using a single material, that is, 17-4 PH stainless steel can be printed successfully with the same contour and dimension using scan data, providing better load-bearing capability and precision fit (Singh et al. 2023). But if an implant has to be fabricated for a smart property then using conventional materials is not possible. The most suitable materials that generate smart property are shape memory alloys, which can remember their shape and can return to initial state even after being twisted. But by using kirigami-based metastructure this smart property can be generated for fabricated smart dental implants.

5.7 Multimaterial dental implant

Dental implants have recently gained a lot of attention from researchers and suppliers due to their prospective worth. The evaluation of long-term success is not yet complete due to the complicated nature of dental implants. The quality of the bone, the nature of the bone–implant contact, and the distribution of stress at the bone–implant interface all have a role in the longevity and effectiveness of dental implants.

Dental osseointegration, described as the implant's firm attachment to the bone and capacity to withstand occlusal forces, is essential for implant stability. The stress at the interface between the bone and implant has a significant impact on the osseointegration process. The main elements impacting stress distribution at the bone–implant interface are the kind and extent of loading, implant material properties, implant geometry, surface structure, and implant design quality (diameter and length) (Geng et al. 2001). At the bone–implement contact, the value of stress is proportional to cortical bone thickness and inversely proportional to cortical bone elastic modulus (Holmes and Loftus 1997).

The stability of dental implants hinges significantly on the mechanical properties of crown materials, particularly the elastic modulus and fatigue strength. When crowns are crafted from materials with a low modulus of elasticity, they exhibit greater flexibility, diminishing the direct transfer of occlusal forces to the underlying bone. Occlusal forces in teeth refer to the forces that occur when the upper and lower teeth come into contact during biting and chewing. These forces are essential for the breakdown of food into smaller, digestible particles. Simultaneously, high fatigue strength ensures the crown's durability, enabling it to withstand the repetitive stresses of chewing and biting without structural failure. Together, these properties reduce the risk of damage and promote long-term stability in dental implant systems. It has been observed that low elastic modulus crown materials operate as shock absorbers for occlusal impact forces. As a result of their dampening action, the influence of occlusal forces on bone–implant contact is greatly reduced (El-Anwar et al. 2014).

Multimaterial AM integration in dental implants involves the cutting-edge use of 3D printing technology to create customized implant structures using a combination of different materials, such as biocompatible metals for the implant body and ceramics or polymers for the crown. This approach allows for precise, patient-specific implant design, optimized functionality, and improved aesthetics (Fig. 5.12). It streamlines the manufacturing process, reducing costs and assembly steps, ultimately leading to enhanced patient outcomes by closely replicating the natural teeth in form and function.

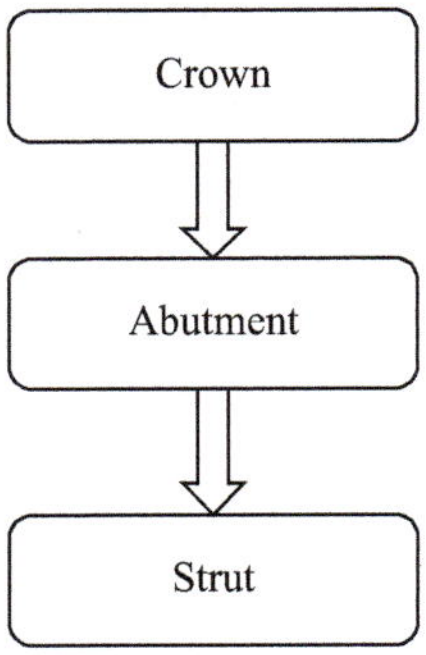

Fig. 5.12: Implant structure from the top.

A functional gradation dental implant design involves tailoring different parts of the implant for specific functions. The top part is strong and gum-friendly to ensure durability and a natural look in the mouth, while the lower section is bone-friendly to promote bonding with the jawbone. This design strategy ensures a stable implant that integrates well with both gum tissue and bone, leading to a successful and long-lasting dental implant.

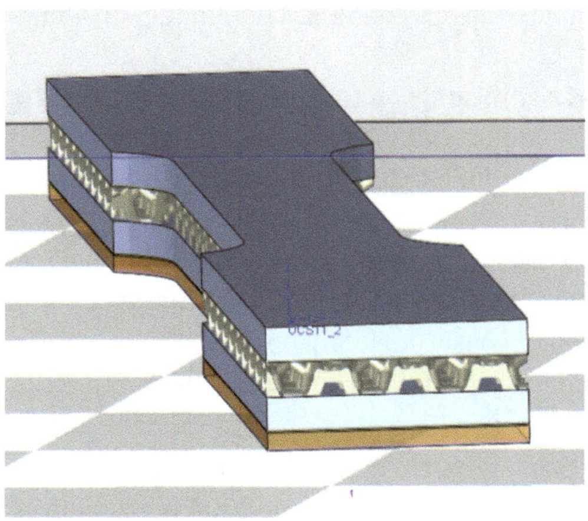

Fig. 5.13: Functional graded lattice infill-based design.

A functional gradation dental implant design can be practically fabricated by laser PBF process-type DMLS. The implants with lower modulus of elasticity exhibit greater flexibility reducing the direct transfer of occlusal forces to the underlying bone. This can be made possible by using lattice infill for strut and abutment parts. With low modulus of elasticity, strut and abutment will act as a shock absorber for occlusal impact forces.

For fabricating multimaterial metastructure-based specimens biocompatible powder with grain size 15–45 µm was filtered in a sieve shaker at a frequency of 60 Hz for

2 h. To remove any unwanted moisture content the powder was heated in an oven for 3 h at 110 °C. The CAD file prepared on SolidWorks was imported to 3dxpert software for creating a support structure, checking printability, stress analysis, part slicing, etc. (Fig. 5.13) Finally, the .FAB3 file of the specimen was created, which is the file format supported by the metal 3D printer. A functionally graded material with lattice infill metastructure can be fabricated successfully by using a single hopper-based DMLS process. The in vitro analysis (Fig. 5.14) proposes that the corrosion rate of fabricated metastructure-based specimens was less than that of virgin single material (Kalia et al. 2023).

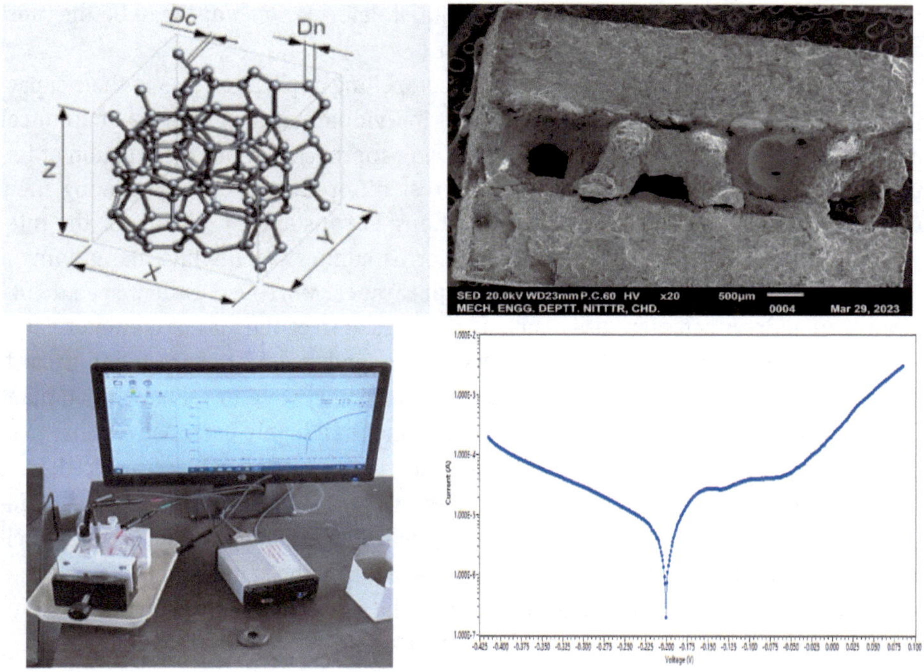

Fig. 5.14: Metastructure-based multimaterial specimen SEM image and tafel plot (Kalia et al. 2023).

Hence, by using multimaterial metastructure for implant fabrication, the limitations of not being able to generate smart property using single material can be overcome. Instead of shape memory alloys, metastructures like kirigami can be employed for the same purpose.

5.8 Conclusions

Making a dental implant out of a single material enables perfect replication of surface contours and dimensions determined straight from scan data. This method ensures adequate dimensional stability, desirable surface roughness, and careful management of porosity percentages, all of which help to the repair of all dental roots, thereby improving load-bearing capacity. The single-material fabricated implant also promotes secondary stability via better osseointegration. When it comes to creating a single-material implant with smart features for health monitoring, traditional materials fall short. Shape memory alloys, which have the extraordinary ability to maintain and return to their former shape even after substantial deformation, appear to be the most appropriate choice in this specialized context.

Multimaterial AM implants represent a compelling application due to their capacity to precisely tailor implant designs to an individual patient's unique anatomical and clinical needs. This advanced manufacturing approach enables the creation of implants with varying material properties across different sections, optimizing their compatibility with specific tissues and functions. For instance, it allows for the integration of materials that are bone-friendly for osseointegration while maintaining a gum-compatible and aesthetically pleasing appearance. Moreover, patient-specific implants can address complex cases, such as anatomical irregularities or trauma-related injuries, by customizing the implant's shape, size, and material composition accordingly. As a result, multimaterial AM offers a promising avenue for enhancing implant functionality, comfort, and overall patient outcomes in a highly personalized manner.

Using a multimaterial metastructure in implant production can assist in reducing the constraints associated with a single material's inability to impart smart capabilities. Rather than relying entirely on shape memory alloys, alternate techniques such as kirigami-based metastructures can be used to achieve the same result.

A functional gradation dental implant with lattice infill produced through multimaterial direct metal printing AM represents an advanced approach to dental implant design and fabrication. This process involves creating an implant structure with varying material properties across its structure. The top part of the implant is engineered for strength and compatibility with the gums, providing durability and a natural appearance. Meanwhile, lattice infill structures within the implant enhance its mechanical properties, such as reducing weight while maintaining strength. This innovative technique also allows for precise control of material composition, which can promote osseointegration with the surrounding bone. Overall, it offers a highly customized and optimized implant solution that blends strength, aesthetics, and integration potential for improved patient outcomes and long-term success.

References

Lukić M, Clarke J, Tuck C, Whittow W, Wells G. Printability of elastomer latex for additive manufacturing or 3D printing. Journal of Applied Polymer Science. 2016 Jan 20;133(4).

Vukicevic M, Mosadegh B, Min JK, Little SH. Cardiac 3D printing and its future directions. JACC: Cardiovascular Imaging. 2017 Feb;10(2):171–84.

Lin L, Fang Y, Liao Y, Chen G, Gao C, Zhu P. 3D printing and digital processing techniques in dentistry: A review of literature. Advanced Engineering Materials. 2019 Jun;21(6):1801013.

Alharbi N, Alharbi S, Cuijpers VM, Osman RB, Wismeijer D. Three-dimensional evaluation of marginal and internal fit of 3D-printed interim restorations fabricated on different finish line designs. Journal of Prosthodontic Research. 2018;62(2):218–26.

Holzmann P, Breitenecker RJ, Soomro AA, Schwarz EJ. User entrepreneur business models in 3D printing. Journal of Manufacturing Technology Management. 2017 Feb 6;28(1):75–94.

Gross BC, Erkal JL, Lockwood SY, Chen C, Spence DM. Evaluation of 3D printing and its potential impact on biotechnology and the chemical sciences.

Sames WJ, List FA, Pannala S, Dehoff RR, Babu SS. The metallurgy and processing science of metal additive manufacturing. International Materials Reviews. 2016 Jul 3;61(5):315–60.

MacDonald E, Wicker R. Multiprocess 3D printing for increasing component functionality. Science. 2016 Sep 30;353(6307):aaf2093.

Smith CJ, Tammas-Williams S, Mahoney PS, Todd I. 3D printing a jet engine: An undergraduate project to exploit additive manufacturing now and in the future. Materials Today Communications. 2018 Sep 1;16:22–25.

Bose S, Ke D, Sahasrabudhe H, Bandyopadhyay A. Additive manufacturing of biomaterials. Progress in Materials Science. 2018 Apr 1;93:45–111.

Harun WS, Kamariah MS, Muhamad N, Ghani SA, Ahmad F, Mohamed Z. A review of powder additive manufacturing processes for metallic biomaterials. Powder Technology. 2018 Mar 1;327:128–51.

Zhang Y, Bandyopadhyay A. Direct fabrication of bimetallic Ti6Al4V+ Al12Si structures via additive manufacturing. Additive Manufacturing. 2019 Oct 1;29:100783.

Zhang Y, Bandyopadhyay A. Direct fabrication of compositionally graded Ti-Al2O3 multi-material structures using Laser Engineered Net Shaping. Additive Manufacturing. 2018 May 1;21:104–11.

Bandyopadhyay A, Heer B. Additive manufacturing of multi-material structures. Materials Science and Engineering: R: Reports. 2018 Jul 1;129:1–6.

Rao BS, Rao TB. Mechanical and tribological properties of 3D printed Al-Si alloys and composites: A review. Silicon. 2021 Sep:1–32.

Vollertsen F. Mechanisms and models for laser forming. Proceedings of the LANE'94. 1994;1:345.

Keshavarzkermani A, Sadowski M, Ladani L. Direct metal laser melting of Inconel 718: Process impact on grain formation and orientation. Journal of Alloys and Compounds. 2018 Mar 5;736:297–305.

Zhang Y, Zhang J. Finite element simulation and experimental validation of distortion and cracking failure phenomena in direct metal laser sintering fabricated component. Additive Manufacturing. 2017 Aug 1;16:49–57.

Sanz C, Navas VG. Structural integrity of direct metal laser sintered parts subjected to thermal and finishing treatments. Journal of Materials Processing Technology. 2013 Dec 1;213(12):2126–36.

Zhao X, Iyer A, Promoppatum P, Yao SC. Numerical modeling of the thermal behavior and residual stress in the direct metal laser sintering process of titanium alloy products. Additive Manufacturing. 2017 Mar 1;14:126–36.

Gajera HM, Dave KG, Darji VP, Abhishek K. Optimization of process parameters of direct metal laser sintering process using fuzzy-based desirability function approach. Journal of the Brazilian Society of Mechanical Sciences and Engineering. 2019 Mar;41:1–23.

Mullen L, Stamp RC, Brooks WK, Jones E, Sutcliffe CJ. Selective laser melting: A regular unit cell approach for the manufacture of porous, titanium, bone in-growth constructs, suitable for orthopedic applications. Journal of Biomedical Materials Research Part B: Applied Biomaterials: An Official Journal of the Society for Biomaterials, the Japanese Society for Biomaterials, and the Australian Society for Biomaterials, and the Korean Society for Biomaterials. 2009 May;89(2):325–34.

Stamp R, Fox P, O'Neill W, Jones E, Sutcliffe C. The development of a scanning strategy for the manufacture of porous biomaterials by selective laser melting. Journal of Materials Science: Materials in Medicine. 2009 Sep;20:1839–48.

Hollander DA, Von Walter M, Wirtz T, Sellei R, Schmidt-Rohlfing B, Paar O, Erli HJ. Structural, mechanical, and in vitro characterization of individually structured Ti–6Al–4V produced by direct laser forming. Biomaterials. 2006 Mar 1;27(7):955–63.

Mangano C, Raspanti M, Traini T, Piattelli A, Sammons R. Stereo imaging and cytocompatibility of a model dental implant surface formed by direct laser fabrication. Journal of Biomedical Materials Research Part A: An Official Journal of the Society for Biomaterials, the Japanese Society for Biomaterials, and the Australian Society for Biomaterials, and the Korean Society for Biomaterials. 2009 Mar 1;88 (3):823–31.

Dabrowski B, Swieszkowski W, Godlinski D, Kurzydlowski KJ. Highly porous titanium scaffolds for orthopaedic applications. Journal of Biomedical Materials Research Part B: Applied Biomaterials. 2010 Oct;95(1):53–61.

Fukuda A, Takemoto M, Saito T, Fujibayashi S, Neo M, Pattanayak DK, Matsushita T, Sasaki K, Nishida N, Kokubo T, Nakamura T. Osteoinduction of porous Ti implants with a channel structure fabricated by selective laser melting. Acta Biomaterialia. 2011 May 1;7(5):2327–36.

Traini T, Mangano C, Sammons RL, Mangano F, Macchi A, Piattelli A. Direct laser metal sintering as a new approach to fabrication of an isoelastic functionally graded material for manufacture of porous titanium dental implants. Dental Materials. 2008 Nov 1;24(11):1525–33.

Inagaki I, Takechi T, Shirai Y, Ariyasu N. Application and features of titanium for the aerospace industry. Nippon Steel & Sumitomo Metal Technical Report. 2014 Jul 22;106(106):22–27.

Boyer RR. An overview on the use of titanium in the aerospace industry. Materials Science and Engineering: A. 1996 Aug 15;213(1–2):103–14.

Niinomi M, Nakai M, Hieda J. Development of new metallic alloys for biomedical applications. Acta Biomaterialia. 2012 Nov 1;8(11):3888–903.

Hsiao CN, Chiou CS, Yang JR. Aging reactions in a 17–4 PH stainless steel. Materials Chemistry and Physics. 2002 Mar 1;74(2):134–42.

Viswanathan UK, Banerjee S, Krishnan R. Effects of aging on the microstructure of 17–4 PH stainless steel. Materials Science and Engineering: A. 1988 Oct 1;104:181–89.

Schade C. Processing, microstructures and properties of a dual phase precipitation-hardening PM stainless steel (Doctoral dissertation, Drexel University).

Kudzal A, McWilliams B, Hofmeister C, Kellogg F, Yu J, Taggart-Scarff J, Liang J. Effect of scan pattern on the microstructure and mechanical properties of powder bed fusion additive manufactured 17–4 stainless steel. Materials & Design. 2017 Nov 5;133:205–15.

Zhang X, Xu H, Li Z, Dong A, Du D, Lei L, Zhang G, Wang D, Zhu G, Sun B. Effect of the scanning strategy on microstructure and mechanical anisotropy of Hastelloy X superalloy produced by Laser Powder Bed Fusion. Materials Characterization. 2021 Mar 1;173:110951.

Ameer MA, Khamis E, Al-Motlaq M. Electrochemical behaviour of recasting Ni–Cr, and Co–Cr non-precious dental alloys. Corrosion Science. 2004 Nov 1;46(11):2825–36.

Gaytan SM, Murr LE, Ramirez DA, Machado BI, Martinez E, Hernandez DH, Martinez JL, Medina F, Wicker RB. A TEM study of cobalt-base alloy prototypes fabricated by EBM. Materials Sciences and Applications. 2011 May 5;2(05):355.

Xin XZ, Chen J, Xiang N, Wei B. Surface properties and corrosion behavior of Co–Cr alloy fabricated with selective laser melting technique. Cell Biochemistry and Biophysics. 2013 Dec;67:983–90.

Monroy K, Delgado J, Ciurana J. Study of the pore formation on CoCrMo alloys by selective laser melting manufacturing process. Procedia Engineering. 2013 Jan 1;63:361–69.

Monroy KP, Delgado J, Sereno L, Ciurana J, Hendrichs NJ. Effects of the selective laser melting manufacturing process on the properties of CoCrMo single tracks. Metals and Materials International. 2014 Sep;20:873–84.

Fu W, Liu S, Jiao J, Xie Z, Huang X, Lu Y, Liu H, Hu S, Zuo E, Kou N, Ma G. Wear resistance and biocompatibility of Co-Cr dental alloys fabricated with CAST and SLM techniques. Materials. 2022 May 2;15(9):3263.

Coffman C, Visser C, Soukup J, Peak M. Crowns and prosthodontics. Wiggs's Veterinary Dentistry: Principles and Practice. 2019 Feb;8:387–410.

Capik I, Ledecký V, Ševčík A. Tooth fracture evaluation and endodontic treatment in dogs. Acta Veterinaria Brno. 2000;69(2):115–22.

Sumanth KB, Srinivasan V. Investigation of laser scan strategy on tensile and microstructural characteristics of Inconel 718 fabricated by DMLS process. Journal of Alloys and Compounds. 2022 May 25;904:164099.

Chang JZ, Tsai PI, Kuo MY, Sun JS, Chen SY, Shen HH. Augmentation of DMLS biomimetic dental implants with weight-bearing strut to balance of biologic and mechanical demands: From bench to animal. Materials. 2019 Jan 7;12(1):164.

Geng JP, Tan KB, Liu GR. Application of finite element analysis in implant dentistry: A review of the literature. The Journal of Prosthetic Dentistry. 2001 Jun 1;85(6):585–98.

Holmes DC, Loftus JT. Influence of bone quality on stress distribution for endosseous implants. The Journal of Oral Implantology. 1997;23(3):104–11.

El-Anwar MI, El-Mofty MS, Awad AH, El-Sheikh SA, El-Zawahry MM. The effect of using different crown and implant materials on bone stress distribution: A finite element study. Egyptian Journal of Oral and Maxillofacial Surgery. 2014 May 1;5(2):58–64.

Singh R, Das A, Anand A. On 3D printing of customized multi-root dental implants for the strategic tooth of canine by direct metal laser sintering. Rapid Prototyping Journal. 2023 Mar 2;29(3):447–59.

Kalia B, Singh R, Pabla BS, Dureja M. On direct metal laser sintering of functionally graded material with virgin and mixed powder, In Press.

Gurwinder Singh, Rupinder Singh, Minhaz Husain*, Vinay Kumar

6 3D metal printing of partially absorbable smart orthopedic implant

Abstract: In the recent past, several studies have been reported on 3D-printed polymer-based partially absorbable smart orthopedic implants. But hitherto little has been reported on metal-based absorbable implants. This study highlights the fabrication and in vitro testing of partially absorbable metallic 3D-printed implants for bovine. The study reports the corrosion analysis performed on 3D-printed 17-4 precipitation hardening stainless steel and thermoplastic composite (polyvinylidene fluoride–hydroxyapatite–chitosan composite) used as smart partially absorbable implant sensor for online health monitoring of bovine (to control the reoccurrence of diaphragmatic hernia).

Graphical abstract:

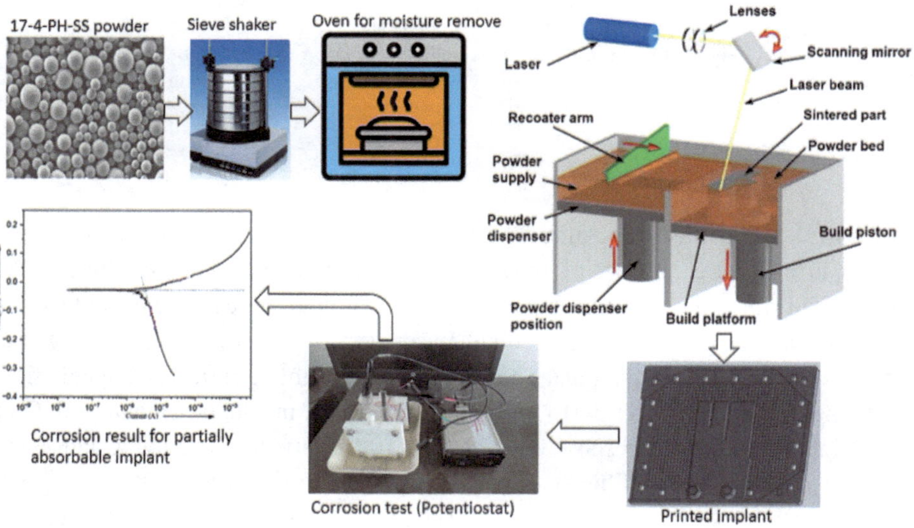

*Corresponding author: Minhaz Husain, Department of Mechanical Engineering, National Institute of Technical Teachers Training and Research, Chandigarh, India, e-mail: minhazme2012@gmail.com
Gurwinder Singh, Rupinder Singh, Department of Mechanical Engineering, National Institute of Technical Teachers Training and Research, Chandigarh, India
Vinay Kumar, University Centre of Research and Development, Chandigarh University, Mohali, Punjab, India

https://doi.org/10.1515/9783111081632-006

Keywords: 3D printing, smart implants, in vitro, metal printing, diaphragmatic hernia, microstrip patch antenna, online health monitoring

6.1 Introduction

Additive manufacturing, also known as 3D printing techniques, builds three-dimensional products layer by layer from computer models. Traditional manufacturing techniques have been revolutionized by this innovative procedure, allowing for quicker production, design flexibility, and customized solutions in a variety of sectors, including aerospace, healthcare, medicine, fashion, and architecture (Singh et al. 2020). The process flowchart of 3D printing is shown in Fig. 6.1.

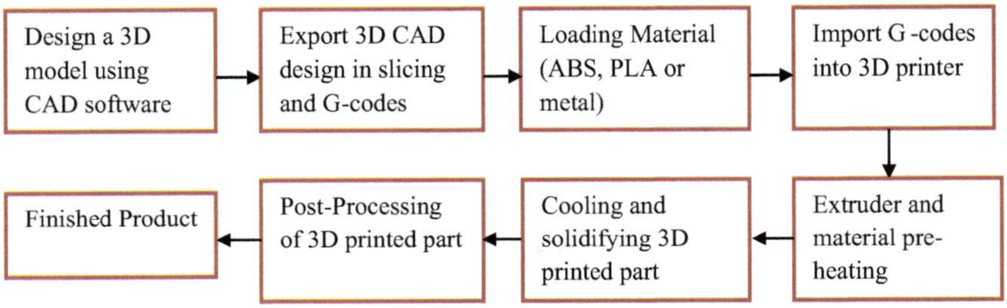

Fig. 6.1: Flow process of 3D printing.

In the area of orthopedic implants, 3D printing has emerged as a game-changing technology, bringing in a new age of individualized and creative healthcare solutions. The use of 3D printing in orthopedic implants has made it possible to create patient-specific solutions that ensure a precise fit and enhance postoperative results (Zamborsky et al. 2019). Wu et al. (2023) highlighted applications of 3D printing in the orthopedic field, including in spine, shoulder, wrist, knee, hip, feet, ankle, maxillofacial, and oral. Complex joint replacement and spiral implants may be made with 3D printing, addressing the individual anatomical difficulties that every patient faces. Furthermore, the capacity to fabricate bone grafts using 3D printing technology has hastened the healing process and improved patient recovery rates (Li et al. 2022). 3D printing in orthopedic implants includes personalized knee and hip replacements, customized cranial implants, and titanium implants. A potential technique for enhancing implant integration with the surrounding bone tissue is the use of trabecular titanium implants. Porous trabecular titanium implants improve implant integration and stability by encouraging improved bone growth. The lattice structures in metal monoblock implants are particularly well suited for 3D printing technology's ability to produce complicated implants with a free-form surface. These implants' interconnected pore surface structure promotes better osteointegration – the process by which the bone naturally grows into the implants, which strengthens the link between the bone and the implant. This method also tackles the problem of stiffness, which can arise with conventional implants and leading to stress-shielding problems, by customizing the porosity and pore size of the implants (Wong 2016). 3D-printed polyvinylidene fluoride (PVDF) thermoplastic composite-based smart partially absorbable orthopedic implant for online health monitoring of canines at industrial, scientific, and medical (ISM) bands was studied (Husain et al. 2023a). Singh et al. (2022) investigated 3D-printed smart sensor polylactic acid (PLA) composite-based orthopedic implant with 17-4 precipitate-hardened (PH) stainless steel (SS) (conductor and ground plane) for online health monitoring of osseointegration at the ISM band. The characteristics of 3D-printed polymer-based implants (along with 4D applications) can closely resemble those of the actual bones, especially in the dentistry and orthopedic fields (Husain et al. 2023b). Wang et al. (2023) investigated 3D-printed (fused deposition modeling) PLA composite (hydroxyapatite (HAp) and lithium (Li)) bone tissue scaffold. A composite biomaterial made of PLA composite exhibits high biocompatibility, degradability, and osteogenic inducibility, and may find use in the healing of fractures in the bone. 3D-printed porous piezoelectric PVDF with graphene oxide (GO) scaffold for bone tissue engineering was fabricated through selective laser sintering (SLS). GO enhances the biocompatibility and mechanical strength, and increases the β-phase of PVDF (Shuai et al. 2020). The use of drug-loaded composite nanofibrous scaffolds with specific characteristics to regulate the stem cell density and encourage bone tissue regeneration was emphasized by (Gao et al. 2016). Drug-loaded composite nanofibrous scaffolds include calcium phosphate mineral, bioactive molecules, and highly organized fiber topology as various cues to resemble bone extracellular matrix. The scaffold was built utilizing nanohydroxyapatite with a surface functionalized with peptides generated from bone morphogenetic protein-7 (Gao et al. 2016). Cao et al. (2020) proposed a 3D-printed Digital light processing (DLP) technique that was used to create porous scaffolds made of ZrO_2/HA composite for use with bone tissue. Strong mechanical properties and strong biocompatibility are benefits of zirconia and HAp, respectively. Gao et al. (2019) fabricated a novel iron–palladium–bredigite (Fe-Pd-bredigite) using the 3D printing technique of selective laser melting (SLM) to enhance the Fe biological activity and degradation behavior in the field of bone/tissue applications. Yang et al. (2020) highlighted a laser-sintering technique used for integrating zeolitic imidazolate framework-8 (ZIF-8), a kind of metal-organic framework, into an already-existing poly-L-lactic acid (PLLA) scaffold. Due to its hydrophobic properties and numerous organic ligands, ZIF-8 developed a favorable interface interaction with PLLA during the procedure. This method considerably increased the scaffold's tensile and compressive strength and enhanced its mechanical characteristics, which can be further used for bone repair applications. Du et al. (2017) proposed pearl/calcium sulfate ($CaSO_4$) scaffolds for bone tissue engineering that have been successfully created via 3D printing with a hydration process. The pearl/$CaSO_4$ scaffolds have a 3D porous structure with constant interconnectivity across the whole scaffold, which improved the compressive strength. In addition, pearl/$CaSO_4$ scaffolds showed improved new bone formation capacity, in comparison to $CaSO_4$ scaffolds, following the implantation of critical-sized rabbit femoral condyle lesions. The applications of 3D printing in orthopedic implants are shown in Fig. 6.2. Table 6.1 shows 3D printing applications in orthopedics with different cases. From the literature, it has been ascertained that previous researchers have investigated 3D-printed polymer-based partially absorbable orthopedic implants with some sensing capabilities but little has been reported on the 3D metal printed partially absorbable smart orthopedic implants.

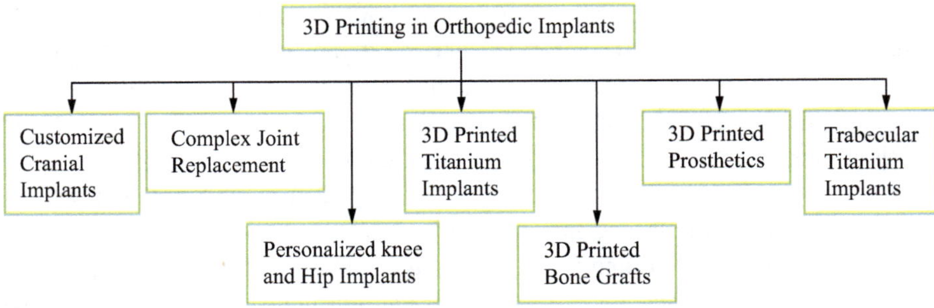

Fig. 6.2: 3D printing in orthopedic applications.

Tab. 6.1: 3D printing in orthopedic applications.

3D printing in orthopedics	Cases	Implant/material type	Printing process
Maxillofacial and oral (Wu et al. 2023)	Complicated craniomaxillofacial reconstruction	Titanium (Ti) plate	Selective laser sintering (SLS)
	Anterior maxillary area (loss of the top front teeth and bone)	Polylactic acid (PLA) for model and Ti implant	Fused deposition modeling (FDM) and SLS
	Severe maxillofacial trauma	Titanium lattice tray (Ti-6Al-4V)	Selective laser melting (SLM)
	Squamous cell carcinoma and mandibular corpus ameloblastoma	Titanium grade II	SLM
	Ameloblastoma, recurrent in the left mandible	Titanium grade V	SLM

Tab. 6.1 (continued)

3D printing in orthopedics	Cases	Implant/material type	Printing process
Shoulder and wrist joint	Total shoulder arthroplasty (TSA) bone problem (Zou et al. 2018)	Shoulder prosthetic with macro porosity (Ti-6Al-4V)	Electron beam machining (EBM)
	Subscapularis muscle synovial sarcoma, extraskeletal ES (Beltrami et al. 2018)	Porous scaffold scapular implant	EBM
	Giant cell tumor of the wrist bone (GCT) (Lu et al. 2018)	Titanium	EBM
	Circular osteosarcoma on the distant side (Chen et al. 2021)	Tantalum porous 3D-printed prosthetic implant	SLS
Knee and hip recovery	Joint-preserving prosthetic repair (Gursan et al. 2021)	3D-printed tibia plate, with hydroxyapatite coating (Ti-6Al-4V)	SLM
	Proximal tibial giant cell tumor (Zhang et al. 2021)	Porous truncated ellipsoid plate (cone type) and porous square strut (frustum type)	EBM
	Severely deformed hip (Wang et al. 2017)	3D-printed hip implant using titanium	EBM
Feet and ankle	Ankle dorsiflexion deficit (Wu et al. 2023)	3D-printed Ti-6Al-4V plate	EBM
	Talar necrosis and collapse (TNC) (Mu et al. 2021)	3D-printed titanium porous talar prosthesis, with articular structures and screw fixation channel	EBM
	Bilateral foot drop (Yao et al. 2021)	Titanium plate implant	EBM
Spine recovery (Wu et al. 2023)	Post-anterior cervical discectomy and fusion (ACDF) symptom recurrence	Porous Ti-6Al-4V cage	SLM
	Rebuilding the cervical lateral majority to keep the neck stable	Ti-6Al-4V extra low interstitials (ELI), 3D-printed lateral mass prosthesis	EBM

6.2 Research gap and problem formulation

Literature reveals that significant studies have been reported on the fabrication of orthopedic implants by 3D printing processes. The Web of Science database outlined 1,195 terms that have been investigated by researchers in respect of fabricating 3D-printed orthopedic implants. Out of 1,195 terms, 45 key research terms were obtained as threshold terms by considering the minimum occurrence of relevant terms as 5. Table 6.2 shows the 45 key terms used to outline the gap in the literature.

Tab. 6.2: List of key terms reported for 3D-printed orthopedic implants.

Id	Term	Occurrences	Relevance scores
1	3D printing	7	0.7519
2	Addition	10	0.2124
3	Additive manufacturing	8	0.5142
4	Alloy	19	0.7431
5	Biocompatibility	26	0.3328
6	Biomaterial	16	0.9386
7	Bone	18	0.1623
8	Cell	17	0.2064
9	Cell proliferation	7	0.4477
10	Coating	32	0.2216
11	Corrosion resistance	7	0.8663
12	Degradation rate	9	1.7874
13	Development	11	0.6655
14	Effect	13	1.3292
15	Fabrication	7	0.7095
16	Implant	60	0.1892
17	Integration	7	0.6673
18	Mechanical property	16	0.4111
19	Metal	11	0.422
20	MPa	11	1.1429
21	Mu m	7	1.5517
22	Orthopedic implant	15	0.312
23	Osteogenesis	5	1.1513
24	PBF	6	2.8684
25	PCL	7	2.709
26	Pore shape	7	3.8837
27	Porosity	13	2.2336
28	Porous Ti6Al4V	8	1.437
29	Scaffold	63	0.7653
30	Selective laser	8	0.3102
31	SLM	5	0.7699
32	Study	25	0.2469
33	Surface	27	1.2078
34	Surface topography	7	4.0101

Tab. 6.2 (continued)

Id	Term	Occurrences	Relevance scores
35	Technology	19	0.1632
36	Ti6Al4V	7	1.0513
37	Time	8	1.0885
38	Titanium	9	0.4406
39	Titanium alloy	5	1.4198
40	TPS	7	2.0962
41	Treatment	8	0.6846
42	Use	9	0.5188
43	Vitro	16	0.1575
44	Vivo	5	0.3714
45	Week	5	0.83

Based on Tab. 6.2, Fig. 6.3 shows the networking diagram of selected research terms, which highlighted that a variety of investigations have been reported on the characterization of mechanical properties, biocompatibility, and fabrication methods to prepare orthopedic implants and scaffolds.

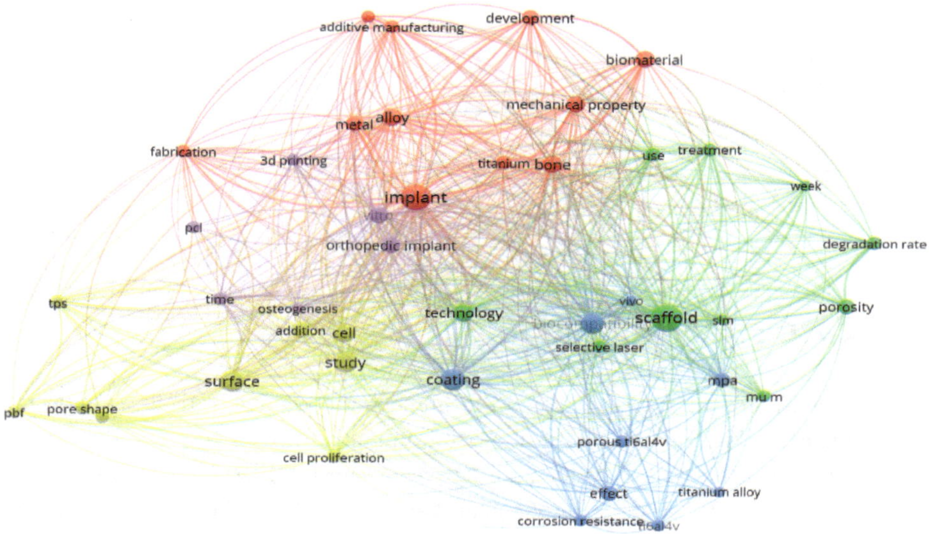

Fig. 6.3: Network diagram of 3D-printed orthopedic implants.

Figures 6.4 and 6.5 outline the gap in the literature of 3D printing explored extensively for biomedical applications and orthopedic implants for humans as well as veterinary animals. But hitherto very few studies have reported on the fabrication of 3D-printed smart orthopedic implants, with partially absorbable properties, for online health monitoring of treated patients. As an extension to the previously reported work on metal-based orthopedic implants, this study reports the investigations performed on in vitro testing of a 3D metal-printed orthopedic implant to ascertain the electrical conductivity-based (dielectric property) sensing capabilities in the partially absorbable implant for smart sensing application, which may be used to fabricate 3D-printed health monitoring solution for different veterinary patients. The tunable dielectric properties of partially absorbable smart implants are useful for field practitioners to take remedial actions in due course of time.

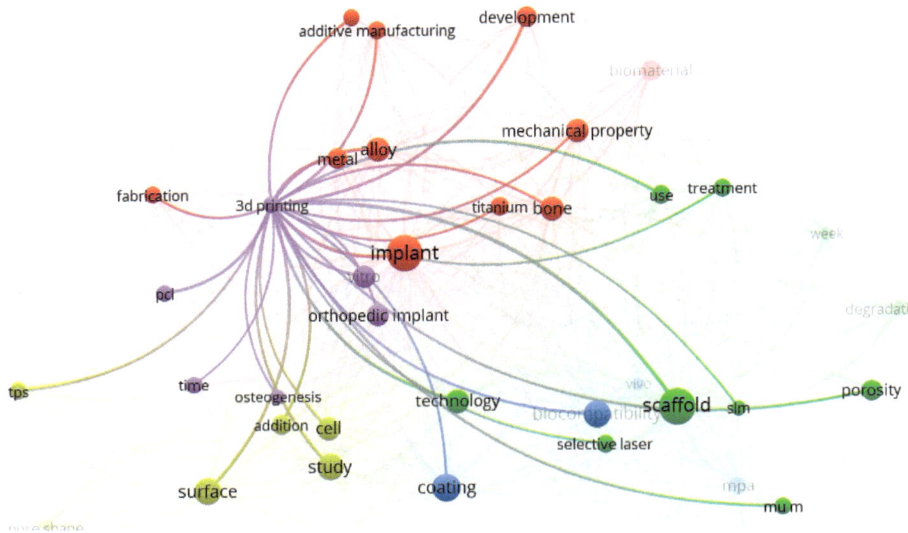

Fig. 6.4: 3D metal-printed orthopedic implants.

6.3 Case study/methodology

In buffaloes, there are four sections of the stomach (i.e., rumen, reticulum, omasum, and abomasum (Fig. 6.6)). As a natural process, initially, the food goes into the rumen. This part is also referred to as food storage, and with the help of an organism, the rumen aids in the digestion of fibrous plant food in buffaloes. Then, it goes to the reticulum. The reticulum helps to sort and prepare the food for digestion. The omasum takes in nutrients and water from the diet and sends the food to the abomasum. The abomasum functions similar to the human stomach by completing the digestive process, so that the buffalo may obtain all the nutrients it requires from its meal. This unique four-part stomach aids the buffalo in the good digestion of its plant-based diet.

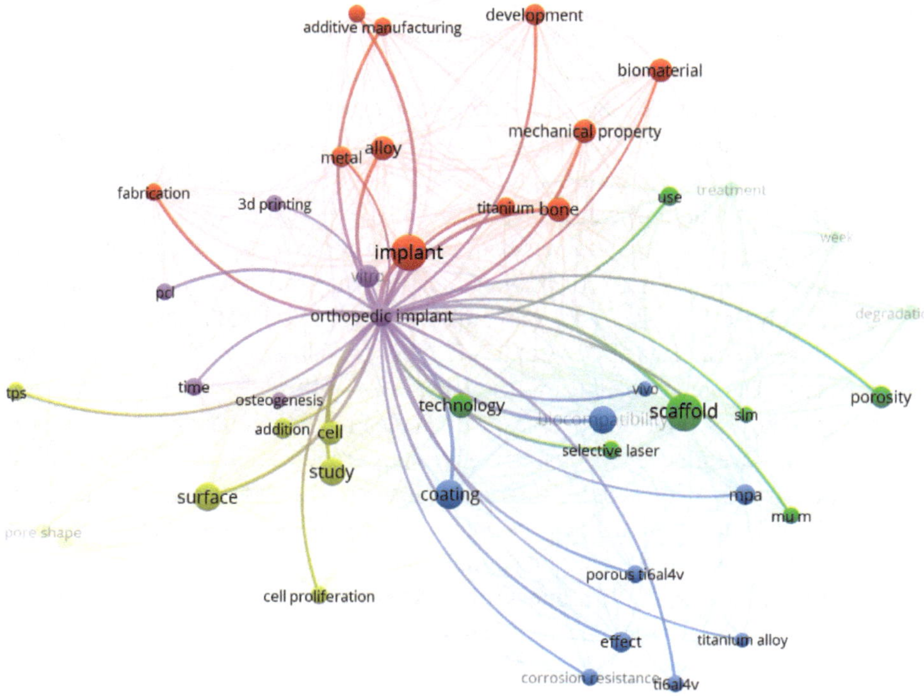

Fig. 6.5: Research gap (orthopedic implants not highlighted with smart partially absorbable keyword).

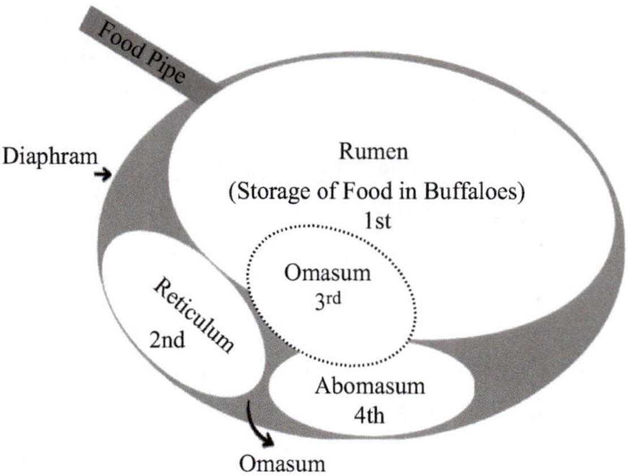

Fig. 6.6: Anatomy of the buffalo stomach.

The buffalo plays a significant role in the community and is integral to both livestock production and the economy. Beyond only supplying food and earnings, it has cultural importance, helping many communities maintain their identity and lifestyle (Nanda and Nakao 2003). In adult milk buffalo, diaphragmatic hernia (DH) is a prevalent thoracoabdominal ailment. In buffalo, increased intra-abdominal pressure, brought due to the viscera's natural weight, different levels of tympany, advanced pregnancy, and the strain of calving, has a significant impact on the development of reticular DH. When these symptoms are present together, DH in mature milk buffalo is more common (Pal and Tayal 2018). Due to a lack of pressure senses such as Gogli tendon organs and pericardiacophrenic arteries in their diaphragm, buffaloes are more likely than cows to develop DH. The stimulation and control that these pressure receptors in cows offer during muscle contraction and movement can help protect the buffaloes against DH. The absence of these pressure receptors in buffaloes may raise the chances of developing DH (Pooniya et al. 2022). Consuming objects like rubber, plastics, leather goods, ropes, and textiles can lead to illness, known as non-metallic foreign body syndrome or silent hazard. The effects of this disease include a significant reduction in milk and meat output, high treatment expenses, probable death, and the possibility of fetal loss in animals that are pregnant. Due to the enormous economic losses caused by this illness, both livestock owners and the agricultural sector risk disastrous outcomes (Abu-Seida and Al-Abbadi 2016). DH is more common in mother's buffalo than in calves or non-pregnant buffaloes. The struggling buffalo may pass away if DH is not treated properly (Talekar et al. 2018). Ultrasonography, radiography, clinical signs, and laparo-rumenotomy are the conventional techniques used to diagnose DH.

6.3.1 Ultrasonography

The process of ultrasonography involves utilizing ultrasonic imaging to see within the animal's reticulum to diagnose DH in buffaloes. High-frequency sound waves or 3.5 MHz and 5 MHz convex transducers are used in the non-invasive diagnostic procedure of ultrasonography, which provides live pictures of the reticulum, rumen, spleen, liver, pleura, heart, organs, tissues, and fluid-filled areas (Abdelaal et al. 2009). The step-by-step procedure to diagnose a buffalo DH using ultrasound is given as under:

- The buffalo is often anesthetized and tethered before the treatment to guarantee its safety and cooperation. Sedation keeps the animal relaxed and minimizes any potential discomfort.
- To improve contact between the ultrasonic probe and the skin, the hair over the region to be studied is cut or shaven. The surface of the skin is covered with a water-based gel, which aids in better sound wave transmission.
- The buffalo is positioned in such a manner that the proper location may be scanned with the ultrasonic probe. The posture of the animal is often standing or lateral recumbent.
- To achieve proper contact, the veterinarian or ultrasound technician places the ultrasound probe on the skin's surface. High-frequency sound waves emitted by the probe travel through the tissues and return, producing pictures on the ultrasound machine.

6.3.2 Radiology

To diagnose DH in buffaloes, radiologists employ X-ray imaging on the right lateral recumbency (Athar et al. 2009) or on the left standing position of buffalo to see the presence of metallic bodies in the reticulum (Pooniya et al. 2022). Images that can be used to spot structural flaws can be produced using X-rays, a kind of electromagnetic radiation that can pierce tissues to variable degrees. Figure 6.7 shows the radiology detection of metallic wire. A step-by-step procedure to diagnose a buffalo DH using radiology is given as under:

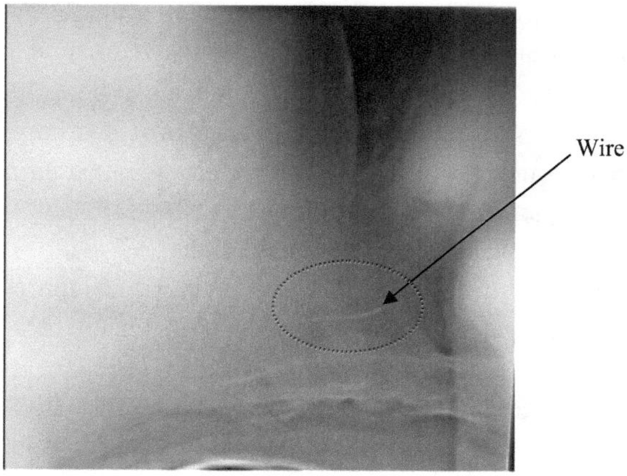

Fig. 6.7: Radiology of the reticulum.

- Sedation is frequently used to relax the buffalo and reduce its tension, to ensure its safety and cooperation throughout the procedure. To immobilize the buffalo and avoid any movement that would impair the quality of the X-ray images, appropriate restraint is also used.
- For the best possible X-ray imaging, the buffalo is positioned in a specific manner. The animal may be placed in a standing, recumbent, or lateral recumbent position, depending on where the DH is thought to be.
- A distance between the X-ray machine and the buffalo is ideal. Depending on the size and density of the animal, the machine's parameters, such as the X-ray energy level and exposure period, are changed.
- By sending X-rays through the buffalo's body, X-ray images are obtained. The density and thickness of the structures encountered determine how much of the X-rays are absorbed or scattered as they move through the tissues. A radiographic image is produced on a specialized X-ray film or a digital detector by recording the pattern of X-ray exposure.

6.3.3 Clinical signs

The majority of animals with a DH typically exhibit tympany, brisket oedema, anorexia, persistent cough, difficulty in breathing, intermittent constipation, or diarrhea with tiny amounts of loose or sparse pasty feces (Pal and Tayal 2018).

6.3.4 Laparo-rumenotomy

A surgery called laparo-rumenotomy is carried out on buffaloes in a standing position to diagnose a DH. It includes using a laparoscopic or rumenotomy method to get access to the abdominal cavity and the herniated organs. In this process, the rumen was cleaned. Subsequently, metallic foreign bodies were found and removed in both the rumen and reticulum (Talekar et al. 2018). A summary of the process is provided here:

- To guarantee that the buffalo is completely asleep and experiences no pain throughout the procedure, general anesthesia is administered. Dorsal recumbency or resting on its back, allows for easier access to the abdomen.
- The abdominal wall of the buffalo is cut into multiple tiny incisions if laparoscopic surgery is chosen. These cuts are used to generate access ports for the laparoscope and surgical tools using trocars (hollow tubes).
- One of the trocars is used to introduce a laparoscope – a long, thin tool with a light and camera connected to it. This enables the vet to see inside the abdomen and spot the DH.

From the above discussion, it is concluded that the veterinarian has limited tools to diagnose DH in buffalo. The success rate of diagnosing DH is 85.18% using ultrasonography, and 66.67% using radiology. Sometimes these methods are unable to detect the nonmetallic foreign objects (plastic, rubber, rope) inside the reticulum, and they remain inside the body of the buffalo or cow. Due to this, the buffalo dies, and mostly several deaths are seen during pregnancy time. So, an innovative Internet of things (IoT) solution – 3D-printed hernia mesh-type implantable sensor – was proposed to overcome the problem of the reoccurrence of DH in bovines. This IoT-based sensor was placed between the rumen and the reticulum. The sensor was based on the microstrip patch antenna (MPA) concept. A typical form of flat, low-profile antenna used in modern communication systems is the microstrip patch antenna. It comprises a dielectric substrate, which is often a thin, rigid board placed on a metallic patch (commonly composed of copper or other conductive materials). The dielectric substrate offers insulation and mechanical support while the metallic patch acts as the radiating element (Kumar et al. 2022). Different geometries of microstrip patch antennas are available and each has a different radiation pattern and set of functional characteristics (Singh and Tripathi 2011). Rectangular and circular patches are a few popular patch forms used in microstrip patch antenna. Rectangular patches are excellent for a wide range of applications because they are straightforward to build and give broadside radiation patterns (Shimu and Ahmed 2016). Microstrip patch antenna is used in a variety of medical and healthcare, including in noninvasive imaging and diagnosis of brain tumors and cancer (Raihan et al. 2017, Ridoy et al. 2021). Fusic et al. 2023 highlighted the early detection and monitoring of breast cancer. Deshmukh and Chorage (2020) outlined continuous and real-time blood glucose level tracking in diabetes. Wearable health monitoring for remote, continuous monitoring of vital signs and general health status (Kumar et al. 2023) and at ISM bands to monitor cardiac functions like heart rate and pacemaker (Hasan et al. 2019) have been reported.

An implantable assembly design made of PVDF polymer and 17-4 PH SS, which serves as a sensor as per the applied Indian patent number: 202311036753., was adopted. The design makes it easy to monitor the recurrence of DH in cattle, after surgery. The sensor resonates on the ISM band in response to pressure from metallic or nonmetallic foreign substances. Figure 6.8 shows the proposed assembly of the sensor in the present case.

Fig. 6.8: Assembly of sensor.

In this innovative idea, 3D printing is utilized to create all the subparts required to prevent the reoccurrence of DH in bovine. In keeping with the concept of the microstrip patch antenna, the mesh is made of PVDF, fabricated through the fused filament fabrication (FFF) process, while the patch and the ground are made of 17-4PH SS, fabricated through direct metal laser sintering (DMLS) (Indian patent application no.: 202311036753). During surgery, the sensor was put on the diaphragm outside the reticulum. The PVDF substrate changes from alpha to beta phase and ultimately modifies the material's dielectric constant when metallic or nonmetallic foreign substances press the sensor. An electronic device may pick up this signal.

Corrosion testing of the 3D-printed hernia mesh implant sensor was carried out, based on the Phadke corrosion testing apparatus. Figure 6.9 shows the methodology of the implant sensor. Normal corrosion happens at a rate set by the equilibrium of opposing electrochemical processes. There is the anodic reaction, during which a metal is oxidized, allowing the metal to release electrons, whereas the other cathodic process removes electrons from the metal by reducing a solution species (typically O_2 or H^+).

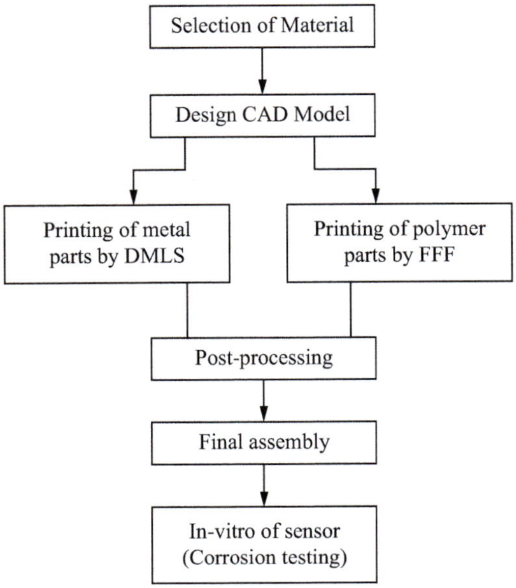

Fig. 6.9: Proposed methodology.

In the potentiostat setup (Fig. 6.10), the counter electrode, the reference electrode, the working sample, and floating ground instruments are used.

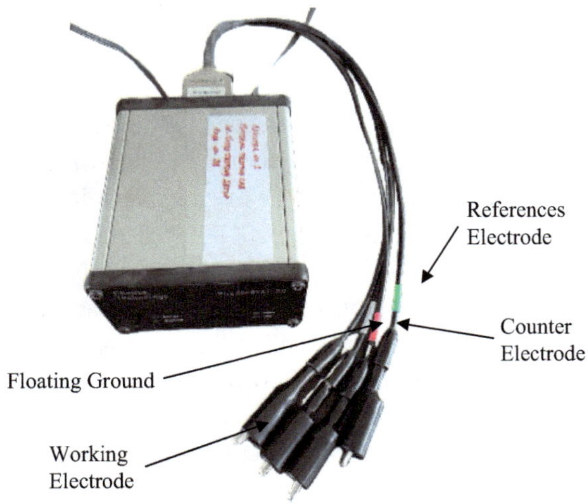

Fig. 6.10: Different electrodes with Phadkestat 20.

6.3.5 Reference electrode

Fig. 6.11: Saturated calomel electrode.

For calculating the potential of the working electrode, the reference electrode acts as a reference point. It offers a reliable electrical potential that is well known and steady, enabling the measurement of the potential of the working electrode. The reference electrode plays a very essential role during corrosion testing. Figure 6.11 shows a saturated calomel electrode (SCE).

6.3.6 Counter electrode

Fig. 6.12: Graphite reference electrode.

The function of the counter electrode (Fig. 6.12) is to complete the electrical circuit during the corrosion testing. To assist the passage of current during the corrosion test, the counter electrode (sometimes referred to as the auxiliary electrode) is utilized in combination with the working electrode. Some other functions of the counter electrode include completing the circuit, maintaining the charge neutrality, and minimizing polarization effects.

6.3.7 Floating ground

The floating electrode (shown in Fig. 6.13) usually refers to a grounding arrangement, where the ground reference is not in direct contact with the Earth's ground or with any other fixed reference point, and the sample to be tested for corrosion is attached to the working electrode.

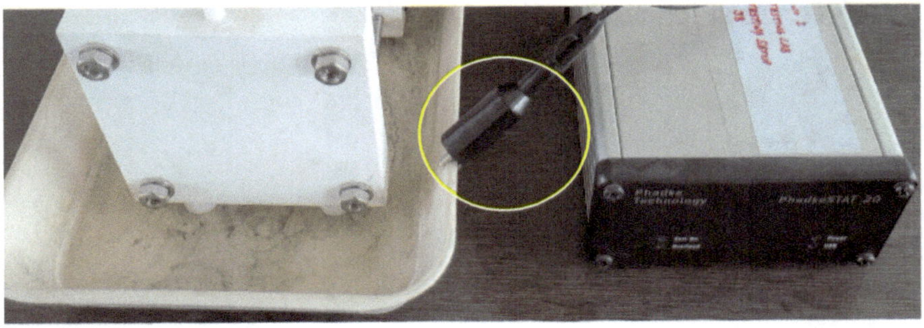

Fig. 6.13: Floating electrode connected with ground.

During the initializing of corrosion testing, the simulating environment was created according to the requirement. For implant applications, the environment/fluid was created according to a particular body. Generally, the most commonly used environmental solution was 7 g sodium chloride (NaCl) with 200 mL water, shown in Fig. 6.14 (a, b). Each sample was attached to the individual electrodes separately for corrosion testing, shown in Fig. 6.14 (c).

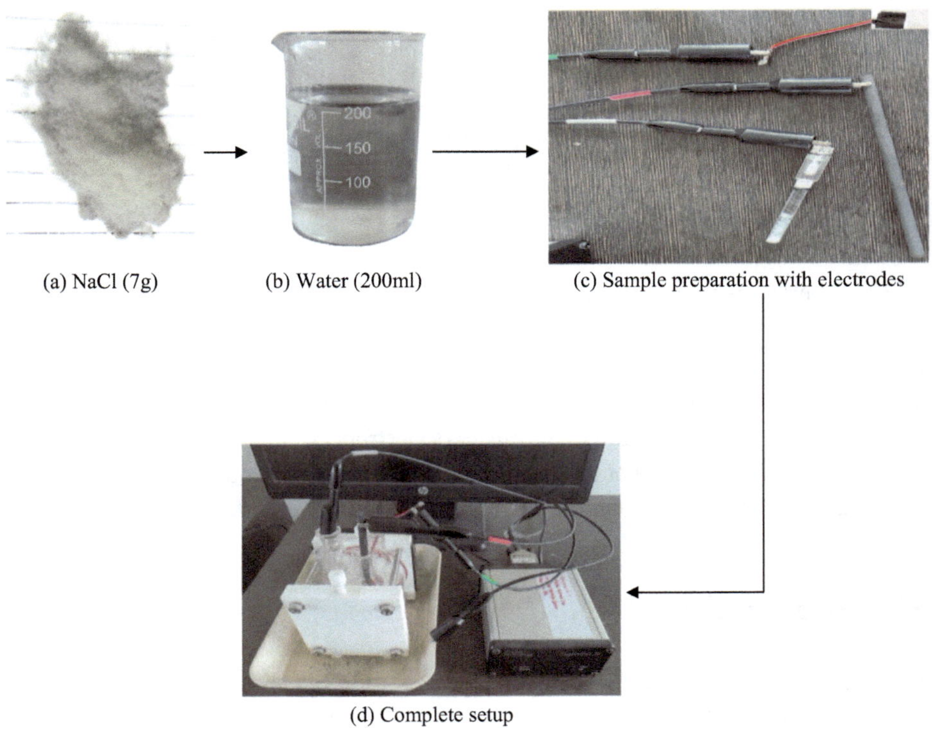

Fig. 6.14: (a) 7 g of NaCl, (b) water 200 mL, (c) preparation samples with each electrode, and (d) complete setup of the in vitro process.

The corrosion test of a 3D-printed hernia mesh-based implant sensor was carried out using 7 g NaCl with 200 mL water. The following steps were followed during the testing:

6.3.8 Open circuit potential (E_{oc})

Open circuit potential (E_{oc}) was carried out. The E_{oc} is the equilibrium potential that the metal assumes when there are no electrical connections to the metal. The initial stage in the majority of electrochemical corrosion experiments is the measurement of E_{oc}. The words E_{oc} (open circuit potential) and E_{corr} (corrosion potential) are sometimes used interchangeably (Hemkemeier et al. 2022, Instruments 2011). Figure 6.15 shows the OCP of a 3D-printed hernia mesh implanted sensor.

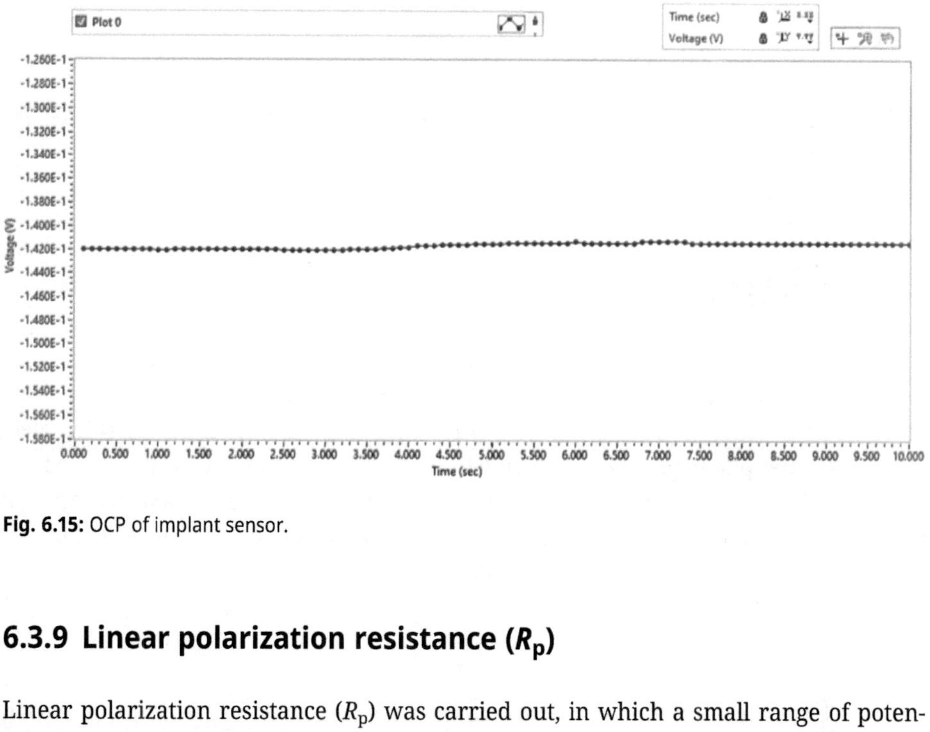

Fig. 6.15: OCP of implant sensor.

6.3.9 Linear polarization resistance (R_p)

Linear polarization resistance (R_p) was carried out, in which a small range of potential that is extremely close to E_{oc} (often ±10 mV) is used to sweep the cell voltage, which results in the current versus voltage curve (Pacheco et al. 2011, Instruments 2011). For measuring the $R_{p,}$ the area of the sample, equivalent weight, and density were calculated. Figure 6.16 shows an R_p graph of the 3D-printed implanted sensor. The R_p was calculated by using the data of the LPR graph exported into an Excel sheet and by the relation of $V = I \times R$, where I is current and R is resistance:

$$I_{cr} = \frac{1}{2.303\ R_p} \times \frac{\beta_a\ \beta_c}{(\beta_a + \beta_c)} \tag{6.1}$$

where I_{cr} is the corrosion current, and $\beta_{a,}$ and β_c are the anodic and cathodic constants, respectively, in equation (6.1).

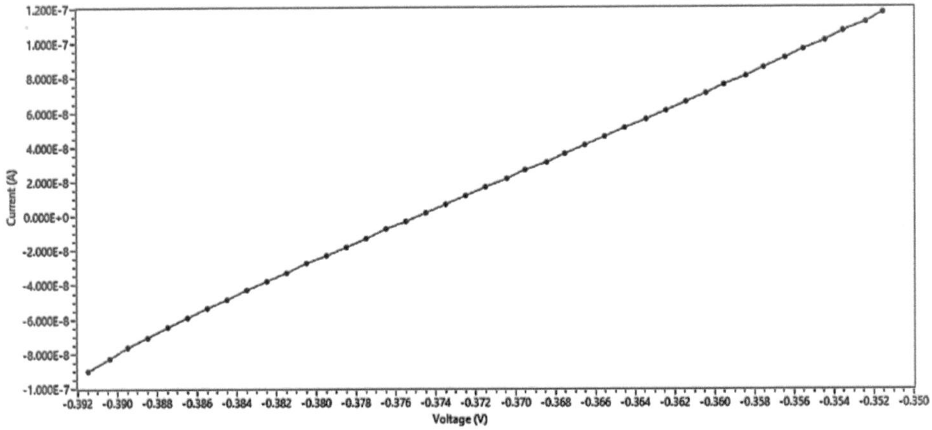

Fig. 6.16: Observed LPR plot.

6.3.10 Tafel plot

A Tafel plot is used to compute the corrosion characteristics, including corrosion rate and corrosion potential as well as corrosion kinetics (McCafferty 2005, Instruments 2011). By measuring the current density of the applied potential for the working electrode, the Tafel plot is produced. Usually, the plot is logarithmic, with the applied potential on the *x*-axis and the current density on the *y*-axis. The Tafel equation explains how a single, isolated reaction behaves. The anodic and cathodic processes in a corrosion system are in opposition to one another:

$$\mathrm{CR} = \frac{I_{cr} \times K \times E_W}{\rho \times A} \tag{6.2}$$

where CR is the corrosion rate, K is the corrosion constant ($K = 3{,}272$, if the corrosion rate is calculated in mm/year, $K = 1.288 \times 10^5$ if the corrosion rate is calculated in milli-inches/year), E_W is equivalent to the sample, ρ is the density of sample g/cm^3, and A is an area of the sample in cm^2.

6.4 Results and discussion

Figure 6.17 shows the Tafel plot (while in vitro analysis) for a complete assembly of sensors (metal and polymer). From the Tafel plot, anodic (β_a) and cathodic constants (β_c) were observed as 0.03758 ± 0.00016 V/decade and 0.45212 ± 0.00012 V/decade, respectively (shown in Fig.6.17). To calculate the corrosion rate, the mathematical equation (6.2) was used.

For calculating the I_{CR}, equation (6.1) was used. Hence, the calculated value of I_{CR} was 0.0000007832 ± 0.0000000011.

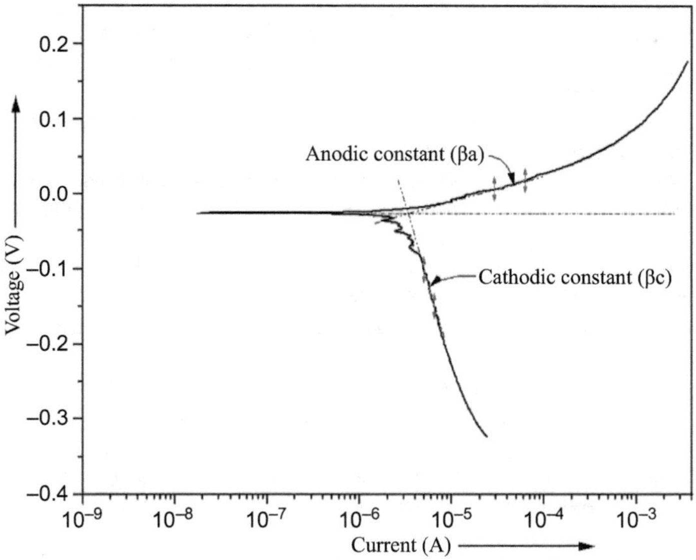

Fig. 6.17: Observed Tafel plot.

Here R_P (linear polarization) was calculated by using the $V = I \times R$ relation. Hence, the calculated value of the corrosion life of 17-4PH SS and PVDF assembly was observed as 0.000311 ± 0.000000001 mm/year.

6.5 Conclusions

In this work, 3D-printed partially absorbable smart metal-based implants have been fabricated using DMLS and FFF techniques. This 3D-printed implant has also 4D applications (metal and polymer) that have been investigated for online health monitoring. The in vitro (corrosion test) performance was also highlighted. In future work, the digital twin of the implant sensor may also be investigated.

References

Abdelaal AM, Floeck M, El Maghawry S, Baumgartner W. Clinical and ultrasonographic differences between cattle and buffaloes with various sequelae of traumatic reticuloperitonitis. Veterinarni Medicina. 2009 Oct 30;54(9):399–406.

Abu-Seida AM, Al-Abbadi OS. Recent advances in the management of foreign body syndrome in cattle and buffaloes: A review. Pakistan Veterinary Journal. 2016 Oct 1;36(4).

Athar H, Mohindroo J, Singh K, Kumar A, Raghunath M. Comparison of radiography and ultrasonography for diagnosis of diaphragmatic hernia in bovines. Veterinary Medicine International. 2010 Oct;2010.

Beltrami G, Ristori G, Scoccianti G, Tamburini A, Capanna R, Campanacci D, Innocenti M. Latissimus dorsi rotational flap combined with a custom-made scapular prosthesis after oncological surgical resection: A report of two patients. BMC Cancer. 2018 Dec;18:1–7.

Cao Y, Shi T, Jiao C, Liang H, Chen R, Tian Z, Zou A, Yang Y, Wei Z, Wang C, Shen L. Fabrication and properties of zirconia/hydroxyapatite composite scaffold based on digital light processing. Ceramics International. 2020 Feb 1;46(2):2300–08.

Chen G, Yin Y, Chen C. Limb-salvage surgery using personalized 3D-printed porous tantalum prosthesis for distal radial osteosarcoma: A case report. Medicine. 2021 Nov 11;100(46).

Deshmukh VV, Chorage SS. Microstrip antennas are used for noninvasive determination of blood glucose levels. In 2020 4th International Conference on Intelligent Computing and Control Systems (ICICCS). 2020 May;13:720–25. IEEE.

Du X, Yu B, Pei P, Ding H, Yu B, Zhu Y. 3D printing of pearl/CaSO 4 composite scaffolds for bone regeneration. Journal of Materials Chemistry B. 2018 6;(3):499–509.

Fusic J, Sugumari T, Sitharthan R. Design of a wearable, flexible Microstrip patch antenna for the detection of breast cancer, 2023.

Gao C, Yao M, Li S, Feng P, Peng S, Shuai C. Highly biodegradable and bioactive Fe-Pd-bredigite biocomposites prepared by selective laser melting. Journal of Advanced Research. 2019 Nov 1;20:91–104.

Gao X, Song J, Zhang Y, Xu X, Zhang S, Ji P, Wei S. Bioinspired design of polycaprolactone composite nanofibers as artificial bone extracellular matrix for bone regeneration application. ACS Applied Materials & Interfaces. 2016 Oct 19;8(41):27594–610.

Gursan O, Celtik M, Yanik B, Husemoglu RB, Havitcioglu H, Gürsan O, CELTIK M, Yanık B. Three-dimensionally-printed joint-preserving prosthetic reconstruction of massive bone defects after malignant tumor resection of the proximal tibia. Cureus. 2021 Mar 9;13(3).

Hasan RR, Rahman MA, Sinha S, Uddin MN, Niloy TS. In-body antenna for monitoring pacemakers. In2019 International Conference on Automation, Computational and Technology Management (ICACTM). 2019 Apr 24;99–102. IEEE.

Hemkemeier TA, Almeida FC, Sales A, Klemm AJ. Corrosion monitoring by open circuit potential in steel reinforcements embedded in cementitious composites with industrial wastes. Case Studies in Construction Materials. 2022 Jun 1;16:e01042.

Husain M, Singh R, Pabla BS. A review on 3D printing of partially absorbable implants. Journal of the Institution of Engineers (India): Series C. 2023a Jul;23:1–20.

Husain M, Singh R, Pabla BS. On PVDF composite as partially absorbable smart implants. Proceedings of the Institution of Mechanical Engineers, Part H: Journal of Engineering in Medicine. 2023b Apr;237(4):517–26.

Indian Patent – a post-surgery detecting device for diaphragmatic hernia in bovines, Application No.: 202311036753, filed on Dated: 27/05/2023.

Instruments G. Getting Started with Electrochemical Corrosion Measurement. Gamry Instruments, Warminster, PA, USA, 2011.

Kumar S, Singh R, Singh AP, Wei Y. Case study for the development of a hybrid composite structure of thermosetting and thermoplastics. In Additive Manufacturing for Plastic Recycling. 2022 Apr 26;141–57. CRC Press.
Kumar S, Singh R, Singh AP, Wei Y. Flexible and wearable patch antenna using additive manufacturing for modern wireless applications. In 3D Printing of Sensors, Actuators, and Antennas for Low-Cost Product Manufacturing. 2023 Feb;10:107–22. CRC Press.
Li Z, Wang Q, Liu G. A review of 3D printed bone implants. Micromachines. 2022 Mar 27;13(4):528.
Lu M, Min L, Xiao C, Li Y, Luo Y, Zhou Y, Zhang W, Tu C. Uncemented three-dimensional-printed prosthetic replacement for giant cell tumor of distal radius: A new design of prosthesis and surgical techniques. Cancer Management and Research. 2018 Feb;8:265–77.
McCafferty E. Validation of corrosion rates measured by the Tafel extrapolation method. Corrosion Science. 2005 Dec 1;47(12):3202–15.
Mu MD, Yang QD, Chen W, Tao X, Zhang CK, Zhang X, Xie MM, Tang KL. Three-dimensional printing talar prostheses for total replacement in talar necrosis and collapse. International Orthopaedics. 2021 Sep;1:1–9.
Nanda AS, Nakao T. Role of buffalo in the socioeconomic development of rural Asia: Current status and future prospectus. Animal Science Journal. 2003 Dec;74(6):443–55.
Pacheco AR, Schokker AJ, Volz JS, Hamilton HR III. Linear polarization resistance tests on corrosion protection degree of post-tensioning grouts. ACI Materials Journal. 2011 Jul 1;108(4).
Pal M, Tayal R. Diagnosis of diaphragmatic hernia in buffaloes: A review. Indian Journal of Veterinary Sciences & Biotechnology. 2018;Apr 8 13(4):78–83.
Pooniya R, Palsania SK, Saini R, Kumar A, Lal M, Kumar H. Surgico-therapeutic management of diaphragmatic hernia in buffalo: A review of 8 cases, 2022.
Raihan R, Bhuiyan MS, Hasan RR, Chowdhury T, Farhin R. A wearable microstrip patch antenna for detecting brain cancer. In2017 IEEE 2nd International Conference on Signal and Image Processing (ICSIP). 2017 Aug 4;432–36. IEEE.
Ridoy PM, Elme KM, Saha P, Hoque MJ, Tulka TK, Rahman MA. Rectangular microstrip patch antenna for biomedical application using ISM band. In2021 International Conference on Intelligent Technologies (CONIT). 2021 Jun 25;1–6. IEEE.
Shimu NJ, Ahmed A. Design and performance analysis of rectangular microstrip patch antenna at 2.45 GHz. In2016 5th International Conference on Informatics, Electronics and Vision (ICIEV). 2016 May;13:1062–66. IEEE.
Shuai C, Zeng Z, Yang Y, Qi F, Peng S, Yang W, He C, Wang G, Qian G. Graphene oxide assists polyvinylidene fluoride scaffold to reconstruct electrical microenvironment of bone tissue. Materials & Design. 2020 May 1;190:108564.
Singh I, Tripathi VS. Micro strip patch antenna and its applications: A survey. International Journal of Computer Technology and Applications. 2011 Sep;2(5):1595–99.
Singh R, Barwar A, Kumar A. Partially absorbable 3D printed innovative orthopedic implants. National Academy Science Letters. 2022 Oct;45(5):417–21.
Singh T, Kumar S, Sehgal S. 3D printing of engineering materials: A state of the art review. Materials Today: Proceedings. 2020 Jan 1;28:1927–31.
Talekar SH, Chaudhary MP, Gamiti KS, Katara KP, Manish S. Surgical management of diaphragmatic hernia in four Jaffarabadi buffaloes and two Gir cows. International Journal of Science, Environment and Technology. 2018;7(2):684–88.
Wang S, Wang L, Liu Y, Ren Y, Jiang L, Li Y, Zhou H, Chen J, Jia W, Li H. 3D printing technology used in severe hip deformity. Experimental and Therapeutic Medicine. 2017 Sep 1;14(3):2595–99.
Wang W, Wei J, Lei D, Wang S, Zhang B, Shang S, Bai B, Zhao C, Zhang W, Zhou C, Zhou H. 3D printing of lithium osteogenic bioactive composite scaffold for enhanced bone regeneration. Composites Part B: Engineering. 2023 May 1;256:110641.

Wong KC. 3D-printed patient-specific applications in orthopedics. Orthopedic Research and Reviews. 2016 Oct;14:57–66.

Wu Y, Liu J, Kang L, Tian J, Zhang X, Hu J, Huang Y, Liu F, Wang H, Wu Z. An overview of 3D printed metal implants in orthopedic applications: Present and future perspectives. Heliyon. 2023Jun 29.

Yang Y, Zan J, Yang W, Qi F, He C, Huang S, Peng S, Shuai C. Metal organic frameworks as a compatible reinforcement in a biopolymer bone scaffold. Materials Chemistry Frontiers. 2020;4(3):973–84.

Yao Y, Mo Z, Wu G, Guo J, Li J, Wang L, Fan Y. A personalized 3D-printed plate for tibiotalocalcaneal arthrodesis: Design, fabrication, biomechanical evaluation and postoperative assessment. Computers in Biology and Medicine. 2021 Jun 1;133:104368.

Zamborsky R, Kilian M, Jacko P, Bernadic M, Hudak R. Perspectives of 3D printing technology in orthopedic surgery. Bratislava Medical Journal/Bratislavske Lekarske Listy. 2019 Jul 1;120(7).

Zhang Y, Lu M, Min L, Wang J, Wang Y, Luo Y, Zhou Y, Duan H, Tu C. Three-dimensional-printed porous implant combined with autograft reconstruction for giant cell tumor in the proximal tibia. Journal of Orthopaedic Surgery and Research. 2021 Dec;16(1):1–9.

Zou Y, Yang Y, Han Q, Yang K, Zhang K, Wang J, Zou Y. Novel exploration of a customized 3D printed shoulder prosthesis in the revision of total shoulder arthroplasty: A case report. Medicine. 2018 Nov;97(47).

Minhaz Husain*, Rupinder Singh, Arvind Kumar, J. Paulo Davim

7 Metastructure-based metal 3D printing for innovative application

Abstract: In the past decade, some studies have been reported on the metastructure-based 3D printing of functional/nonfunctional prototypes. Also, several studies have stated the fabrication of microstrip patch antenna-based sensors using 3D printing. But hitherto little has been reported on the use of metastructure-based design of patches for conformal sensors in biomedical applications. This study highlights the investigation performed on different metastructures and the effect of these metastructures on resonance frequency (R_f) and specific absorption rate (SAR) value of biosensors. The patch and ground plane were printed by direct metal laser sintering (DMLS) process using two different metastructures (octet truss and dodecahedron) for enhancing conformability. The result of this study suggests that the fabrication of a patch with an octet truss metastructure has a better SAR value of 1.576 W/kg (without muscle, skin, and fat) and 1.40 W/kg (with muscle, skin, and fat). The R_f and reflection coefficient (S_{11}) without muscle, skin, and fat were observed as 4.02 GHz and −16.25 dB, and with muscle, skin, and fat was observed as 3.84 GHz and −31.46 dB, respectively for biosensor.

*__Corresponding author: Minhaz Husain,__ Department of Mechanical Engineering, National Institute of Technical Teachers Training and Research, Chandigarh, India,
e-mail: minhazme2012@gmail.com
Rupinder Singh, Arvind Kumar, Department of Mechanical Engineering, National Institute of Technical Teachers Training and Research, Chandigarh, India
J. Paulo Davim, Department of Mechanical Engineering, University of Aveiro, Portugal

https://doi.org/10.1515/9783111081632-007

Graphical abstract:

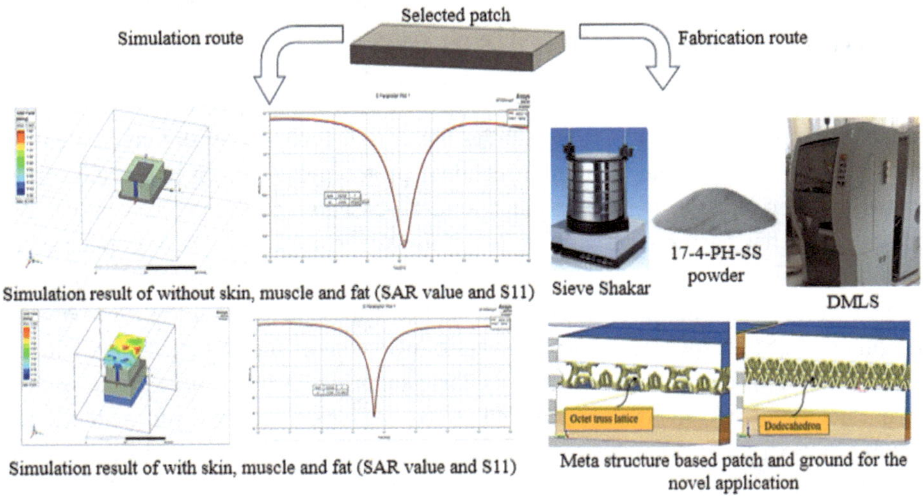

Keywords: Metastructure, biosensor, microstrip patch, DMLS, resonating frequency

7.1 Introduction

The patella (kneecap) of the dog dislocates from its natural position, which could result in discomfort and limping (McLaughlin 2002). Additionally, injuries or fractures to the bones near the knee joint might result in stifle issues. Some breeds are predisposed to stifle joint, developing problems such as hip dysplasia or elbow dysplasia, which can impair mobility and gait (Morgan et al. 2000). The problem can be identified by physical examination, imaging (such as X-rays), and by recommending a suitable course of action, which may involve drugs, surgery, physical therapy, or lifestyle changes.

Canine stifle problems can include a variety of ailments and wounds that affect the knee joint, which is analogous to the human knee (Proffen et al. 2012). It happens when the patella (kneecap) gets displaced from its natural location within the groove of the femur (thigh bone). Congenital (existing from birth) or acquired (acquired later in life) causes like trauma or injury can cause this syndrome. One of the most prevalent stifle issues in canines is an anterior cruciate ligament (ACL) tear. It frequently happens as a result of abrupt knee twisting or overextension and can result in lameness, discomfort, and instability. The canine patella (kneecap) dislocates toward the inside of the knee joint, resulting in temporary or persistent lameness and pain. This condition is known as medial patellar luxation (MPL) (Mostafa 2008). Similar symptoms to those caused by MPL result from lateral patellar luxation, which is a dislocation of the patella toward the outside of the knee joint (Ward et al. 2007). A complete examination by a veterinarian is necessary to identify the precise stifle issue. This assessment frequently includes a physical examination, X-rays, and occasionally more sophisticated imaging methods like magnetic resonance imaging (MRI) or computed tomography (CT) scans.

Depending on the degree of patellar luxation, the canine's patella health, and other considerations established by the veterinarian, either conservative therapy or surgery may be the best course of action (Oshin 2016). Early detection and adequate care can help the canine live a better quality of life and avert chronic joint pain and discomfort., A scale from 1 to 4 is generally used to denote the degree of patellar luxation in dogs. An explanation of the various patellar luxation degrees is provided below (Stein and Schmoekel 2008, Nilsson et al. 2018)

Grade 1: This kind of patellar luxation is the least severe. The patella can be physically luxated (pushed out of its natural position) in Grade 1 instances, but it will quickly snap back into place inside the femoral groove when released. Canines with Grade 1 patellar luxation might not exhibit any clinical symptoms and might not need surgery. Monitoring and conservative care, such as physical therapy and joint supplements, may be advised.

Grade 2: This condition occurs when the patella voluntarily slips out of place while being moved or being examined physically. The patella may remain out of the groove after moving out of position until being physically forced back into it. Canines with Grade 2 patellar luxation frequently display intermittent lameness; thus surgical treatment may be needed to stabilize the patella.

Grade 3: The patella is frequently displaced in this grade, and it can only be physically repositioned into the femoral groove. Grade 3 patellar luxation in dogs frequently results in ongoing discomfort and lameness. Surgery is frequently advised to maintain the patella in place and adjust the structures around the knee joint.

Grade 4: In this grade, patellar luxation is of the most severe kind. The patella is irreversibly displaced and is unable to be manually placed back in the femoral groove in Grade 4 instances. Canines experiencing Grade 4 patellar luxation frequently have severe and persistent lameness and are more likely to experience side effects including osteoarthritis. Almost usually, surgical intervention is required to treat the problem.

Animal implants must be created from biocompatible substances that will not trigger an immunological reaction in the animal or endanger its well-being (Husain et al. 2023a). Medical-grade silicone, titanium, or polymers are typical building materials. The implant may need a power supply, such as batteries or energy-collecting equipment, depending on its intended use. Sensors are commonly used in animal implants to gather data. These might include, temperature sensors, GPS receivers, or other specialized sensors (Husain et al. 2023b). Data transmission from implants to receivers or other external devices may be possible. Bluetooth, radio frequency identification devices, and other wireless communication protocols can all be used for this (Khaleghi et al. 2010).

The size and shape of the implant should match the anatomy of the particular animal type (Husain et al. 2023c). In general, smaller implants are preferable in order to reduce any discomfort or disruption in the animal's natural behavior. To avoid infections after being implanted in an animal, implants must be sterilized (Mulcahy 2013). For the implant to remain in place and avoid migration or expulsion, it must be firmly anchored inside the animal's body. Some implants could be created with short-term usage (intramedullary pin), while others may be created with long-term retention (hip, shoulder, knee replacement). It is vital to remember that an animal implant's particular design and functionality will change depending on whether it is being used for research, tracking, medicinal treatment, or other applications. Additionally, extreme caution and attention to the welfare of the animals involved should be exercised while using animal implants (Pearce et al. 2007).

Metastructure-based metal 3D printing is a method for producing complex, lattice-like structures with distinctive qualities and uses.(Singh and Husain 2023). These novel structures may be tailored for particular purposes and are created at the micro- or nanoscale. Metastructures may be created to produce robust yet lightweight components for spacecraft and aviation. This technique makes it possible to create parts that are lighter while still being structurally sound (Husain et al. 2023d), which enhances performance and fuel economy. Orthopedics and dentistry can use customized lattice architectures to create implants that closely resemble the mechanical characteristics of human bone or teeth. Better integration and healing for the patients may result from this.

Metastructures can be used to create individualized orthotic and prosthetic devices with improved conformability and mechanical qualities (Brooks et al. 2022). Metastructure-based metal 3D printing opens up new design possibilities for consumer goods including athletic goods, which value lightweight and durable materials (Husain et al. 2023e). A flexible technique, metastructure-based metal 3D printing has the potential to be used in a wide range of sectors. It is possible that as technology develops, even more creative and unique applications will appear, fostering breakthroughs in a variety of disciplines (Yang et al. 2021).

In present times, additive manufacturing (AM) techniques are used for the fabrication of scaffolds, implant sensors, etc. (Shahrubudin et al. 2019). The microstrip patch antenna (MPA) has a bright future because MPA sensors are light in weight and work with two or more frequencies (Rana et al. 2023). MPA sensors are easy to fabricate due to their plane structure (Singh and Tripathi 2011). In the healthcare sector, the MPA sensor is widely used as a wearable sensor. The design of the MPA depends upon the material selection and tissues of the human body (Sarsamba and Yanamshetti 2017). The MPA is an assembly of three parts ground plane (conductor), substrate (insulator), and patch (conductor). The operating frequency of MPA is 2.45 GHz and 50 Ω impedance (Dahri et al. 2018). The DMLS is the an established process of AM for the fabrication of metal-based patches. Commercially several of metastructures are available, but octet truss and dodecahedron meta-structure have good mechanical properties because these metastructures have less area of contact (Nazir et al. 2019).

7.2 Research gap

From literature review and bibliographic analysis, the Web of Science database was explored for the keywords "online health monitoring of implant" from 2003 to 2023, and 57 results were found. These 57 results were saved as a plain text file. The plain text file was further processed in the open-source software VOSviewer. Considering a minimum number of occurrences of a term 4, 66 meet the threshold out of the 2,642 terms. A relevance score was determined for each of the 66 words. The most pertinent terms were chosen based on this score. The default choice was to select the 60% most relevant terms, with 40 terms to be selected based on relevance score (Tab. 7.1). Based on Tab. 7.1, a network diagram is shown in Fig. 7.1. Based on Fig. 7.1, the research gap is shown in Fig. 7.2 for the node of "online health monitoring."

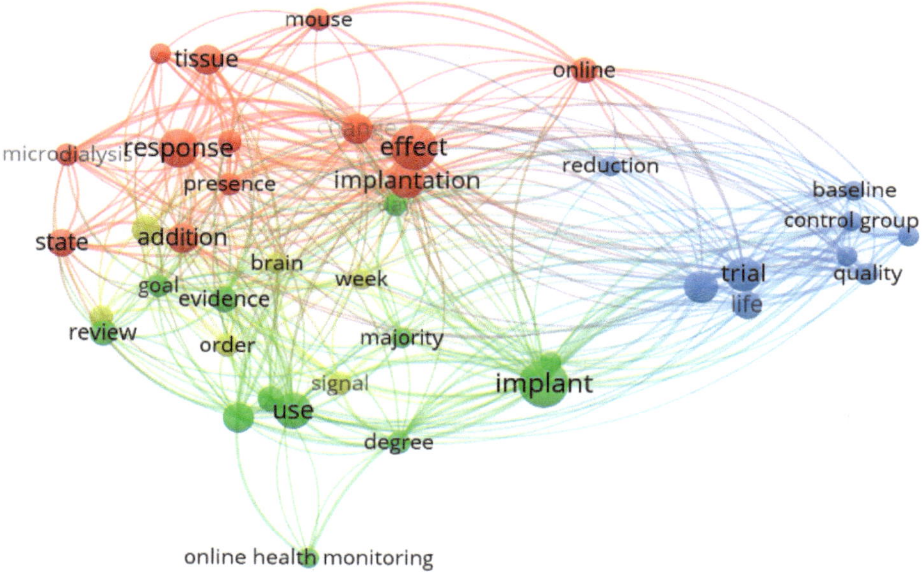

Fig. 7.1: Network diagram for the keywords "online health monitoring of implant."

Similarly, the keywords "metastructure-based patch for antenna" were explored in the Web of Science database from 2003 to 2023, and 2 results were found. These two results were saved as plain text files. The plain text file was further processed in the open-source software VOSviewer. Considering a minimum number of occurrences of

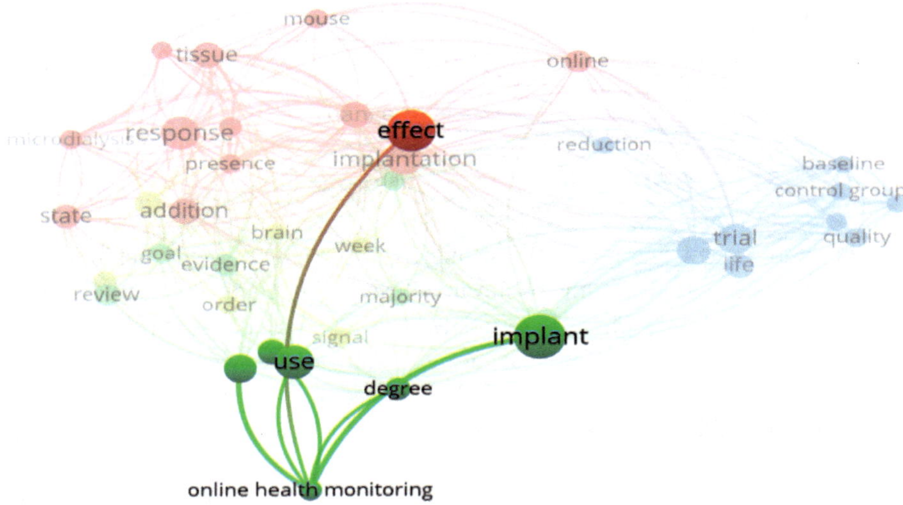

Fig. 7.2: Research gap for the node "online health monitoring."

Tab. 7.1: Occurrence scores for the keywords "online health monitoring."

Id	Term	Occurrences	Relevance scores
1	Addition	8	0.8674
2	Baseline	4	1.4161
3	Brain	4	0.6177
4	Change	8	0.3003
5	Characteristic	8	0.758
6	Combination	5	0.8935
7	Control group	4	1.4099
8	Day	5	0.2919
9	Degree	5	1.0041
10	Effect	15	0.2253
11	Effectiveness	4	1.8378
12	Evidence	6	0.6002
13	Goal	5	0.8287
14	Implant	17	0.6148
15	Implantation	10	0.2482
16	Life	7	1.5678
17	Majority	4	0.4693
18	Measurement	7	1.242
19	Micro dialysis	5	1.6038
20	Microdialysis probe	4	1.6307
21	Mouse	5	1.1715
22	Online	6	0.5913
23	Online health monitoring	4	1.3554
24	Order	5	0.7621

Tab. 7.1 (continued)

Id	Term	Occurrences	Relevance scores
25	Paper	6	0.7095
26	Participant	9	1.1272
27	Presence	5	0.9843
28	Quality	5	1.8499
29	Real time	5	1.3616
30	Reduction	4	0.3533
31	Response	12	0.9169
32	Review	6	1.6399
33	Signal	5	0.8348
34	State	7	1.7494
35	Study protocol	4	1.9475
36	Tissue	8	1.2972
37	Trial	10	1.2733
38	Use	11	0.7031
39	Week	4	0.3916
40	Woman	4	0.5527

a term 1, 129 met the threshold out of the 129 terms. For each of the 129 terms, a relevance score was calculated. Based on this score, the most relevant terms were selected. The default choice was to select the 60% most relevant terms, with 67 terms to be selected based on relevance score (Tab. 7.2). Based on Tab. 7.2, a network diagram is shown in Fig. 7.3. Research gap is shown in Fig. 7.4 for the node of "glass substrate."

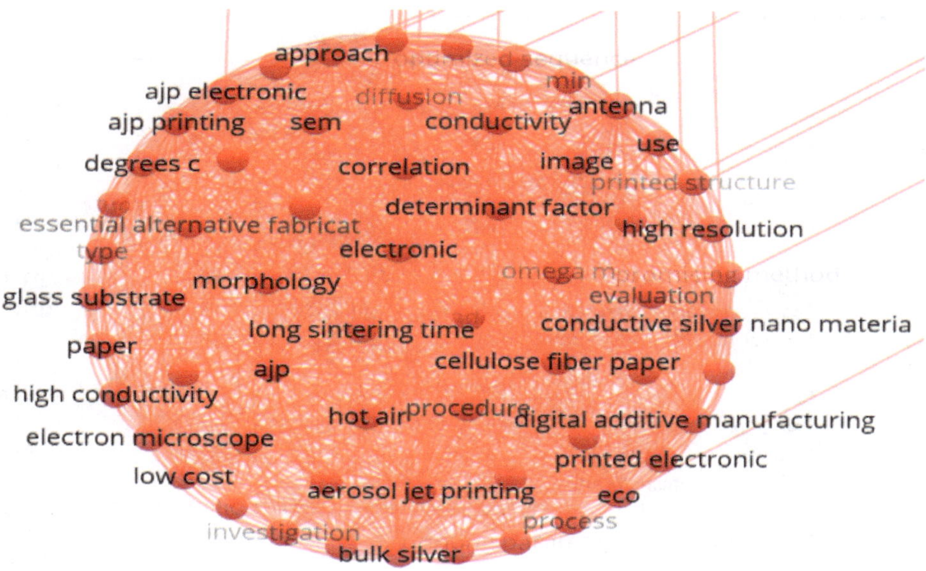

Fig. 7.3: Network diagram for the keyword "metastructure-based substrate for antenna."

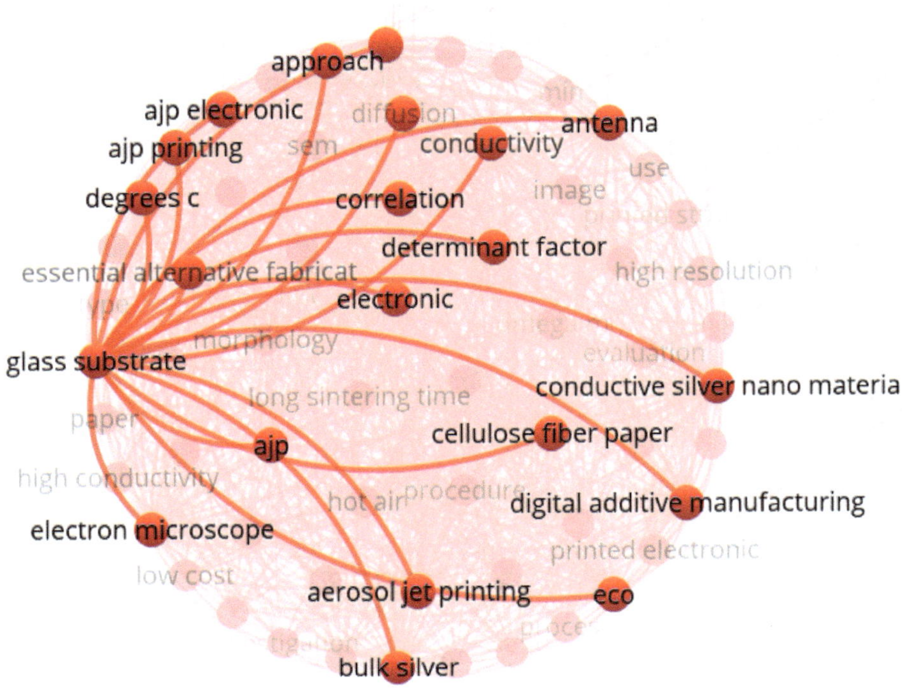

Fig. 7.4: Research gap for the node "glass substrate."

Tab. 7.2: Relevance scores for the keyword "metastructure-based substrate for antenna."

Id	Term	Occurrences	Relevance scores
1	Aerosol jet printing	1	0.0911
2	AJP	1	0.0911
3	AJP electronic	1	0.0911
4	AJP printing	1	0.0911
5	Antenna	1	0.0911
6	Approach	1	0.0911
7	Bulk silver	1	0.0911
8	Cellulose fiber paper	1	0.0911
9	Conductive silver nanomaterial	1	0.0911
10	Conductivity	1	0.0911
11	Correlation	1	0.0911
12	Degrees c	1	0.0911
13	Determinant factor	1	0.0911
14	Diffusion	1	0.0911
15	Digital additive manufacturing	1	0.0911
16	Eco	1	0.0911
17	Electromagnetic functioning surface	1	0.0911
18	Electron microscope	1	0.0911

Tab. 7.2 (continued)

Id	Term	Occurrences	Relevance scores
19	Electronic	1	0.0911
20	Essential alternative fabrication approach	1	0.0911
21	Evaluation	1	0.0911
22	Flexible electronic circuit	1	0.0911
23	Glass substrate	1	0.0911
24	Good adhesion	1	0.0911
25	High conductivity	1	0.0911
26	High resolution	1	0.0911
27	High sintering temperature	1	0.0911
28	Hot air	1	0.0911
29	Image	1	0.0911
30	Investigation	1	0.0911
31	Long sintering time	1	0.0911
32	Low cost	1	0.0911
33	Min	1	0.0911
34	Morphology	1	0.0911
35	New sintering method	1	0.0911
36	Omega m	1	0.0911
37	Optimized sequence	1	0.0911
38	Paper	1	0.0911
39	Paper substrate	1	0.0911
40	Prepared set	1	4.749
41	Printed electronic	1	0.0911
42	Printed metallic structure peel	1	0.0911
43	Printed structure	1	0.0911
44	Printing parameter	1	0.0911
45	Procedure	1	0.0911
46	Process	1	0.0911
47	Promising method	1	0.0911
48	Property	1	0.0911
49	R_f	1	4.749
50	Radiating frequency	1	4.749
51	Radiofrequency	1	4.749
52	Reconfigurable metasurface	1	0.0911
53	Reinforcement	1	4.749
54	Researcher	1	4.749
55	Response	1	4.749
56	Sm	1	0.0911
57	S21	1	4.749
58	Sem	1	0.0911
59	Sensor	1	4.749
60	Sensor development	1	4.749
61	Sensor fabrication	1	4.749
62	Sequence	1	0.0911
63	Set	1	4.749
64	Shape memory effect	1	4.749

Tab. 7.2 (continued)

Id	Term	Occurrences	Relevance scores
65	Sheet resistance	1	0.0911
66	Significant attention	1	4.749
67	Sintering parameter	1	0.0911
68	Sintering temperature	1	0.0911
69	Sintering time	1	0.0911
70	Structure	2	0.2071
71	Study	1	0.0911
72	Suitability	1	4.749
73	Surface	1	0.0911
74	Tremendous flexibility	1	0.0911
75	Type	1	0.0911
76	Use	1	0.0911
77	Window	1	0.0911

From the literature review and bibliometric analysis, veterinarian's clinical examination and different diagnostic instruments and imaging methods, like X-rays, MRIs, and CT scans, are frequently used to monitor the health of a canine stifle joint. Several technologies and gadgets can tangentially help in assessing joint health and tracking changes over time, even though there are no dedicated sensors created specifically for stifle joint health monitoring in the case of canines. To monitor how a canine distributes its weight when standing or walking, pressure-sensitive mats or insoles can be used in their environment. Stifle joint issues may be indicated by altered pressure patterns. To evaluate the joint function, specialized software and sensors may examine a dog's motions and joint angles through various activities.

7.3 Materials and method

Figure 7.5 depicts the methods used for the current investigation.

7.4 Experimentation and discussion

The patch and ground were designed using SolidWorks software. The design and modeling of the sensor were done using high-frequency structure simulator (HFSS) software. The dimension of the patch is 12 × 6 × 0.90 mm as shown in Fig. 7.6.

The simulation of the designed MPA was done on HFSS software. In Fig. 7.7(a) the simulation of the designed MPA without muscle, skin, and fat is shown.

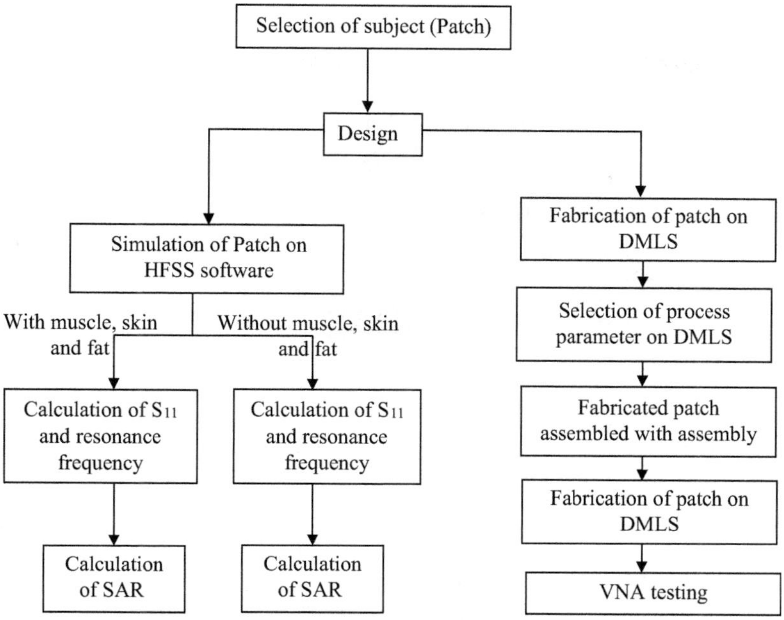

Fig. 7.5: Adopted methodology.

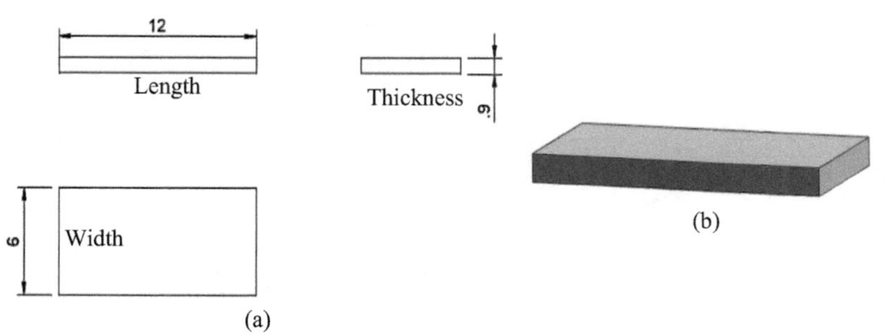

Fig. 7.6: Length, width, and thickness of patch (a), and 3D view of the patch (b).

In Fig. 7.8, the simulation of the designed MPA with the muscle, skin, and fat is shown with SAR values.

7.4.1 Fabrication of patch on DMLS

A powerful laser is used in the DMLS AM technique to selectively fuse metal powder particles, layer by layer, to produce 3D metal objects. In the first step, computer-aided design (CAD) software (SolidWorks) is used to create a 3D digital model (metastructure-based substrate). A collection of 2D digital slices is created by slicing this digital

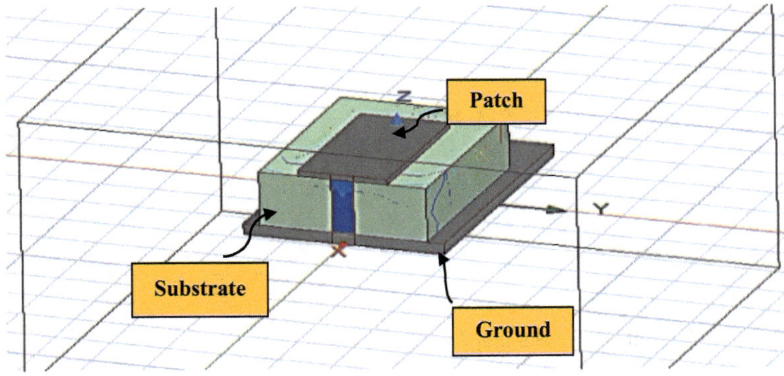

(a)

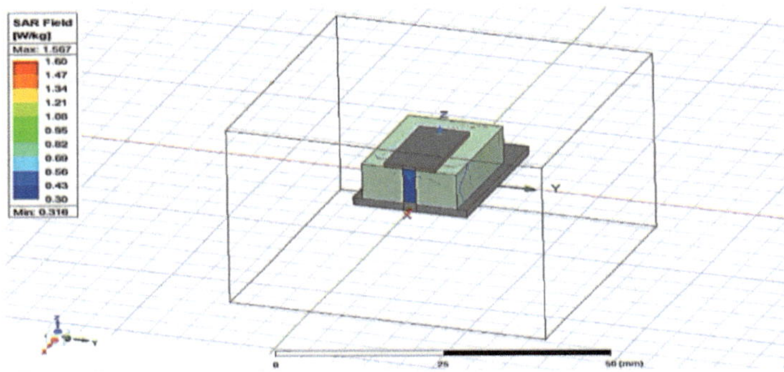

(b)

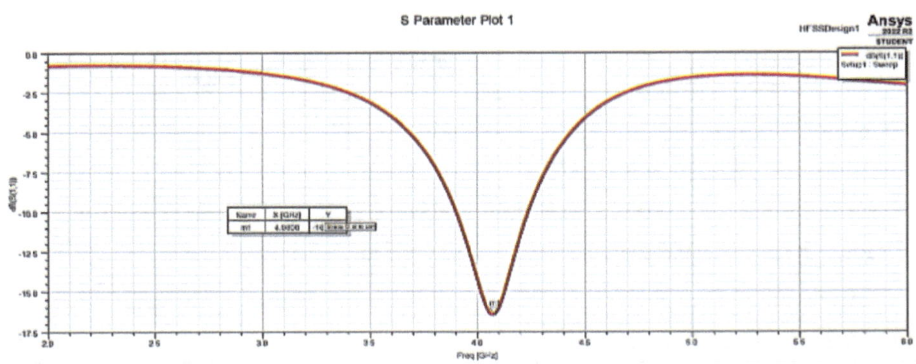

(c)

Fig. 7.7: (a) Simulation of metastructure-based biosensor on the concept of MPA in HFSS, (b) SAR calculation for the designed biosensor, and (c) observed R_f.

model into tiny cross-sectional layers using 3DXpert software. Depending on the mechanical and thermal characteristics needed for the finished item, the 17-4-PH-SS was used to fabricate the patch and ground of the sensors on DMLS. Titanium, aluminum, nickel alloys, stainless steel, and other materials are often utilized in DMLS. The next

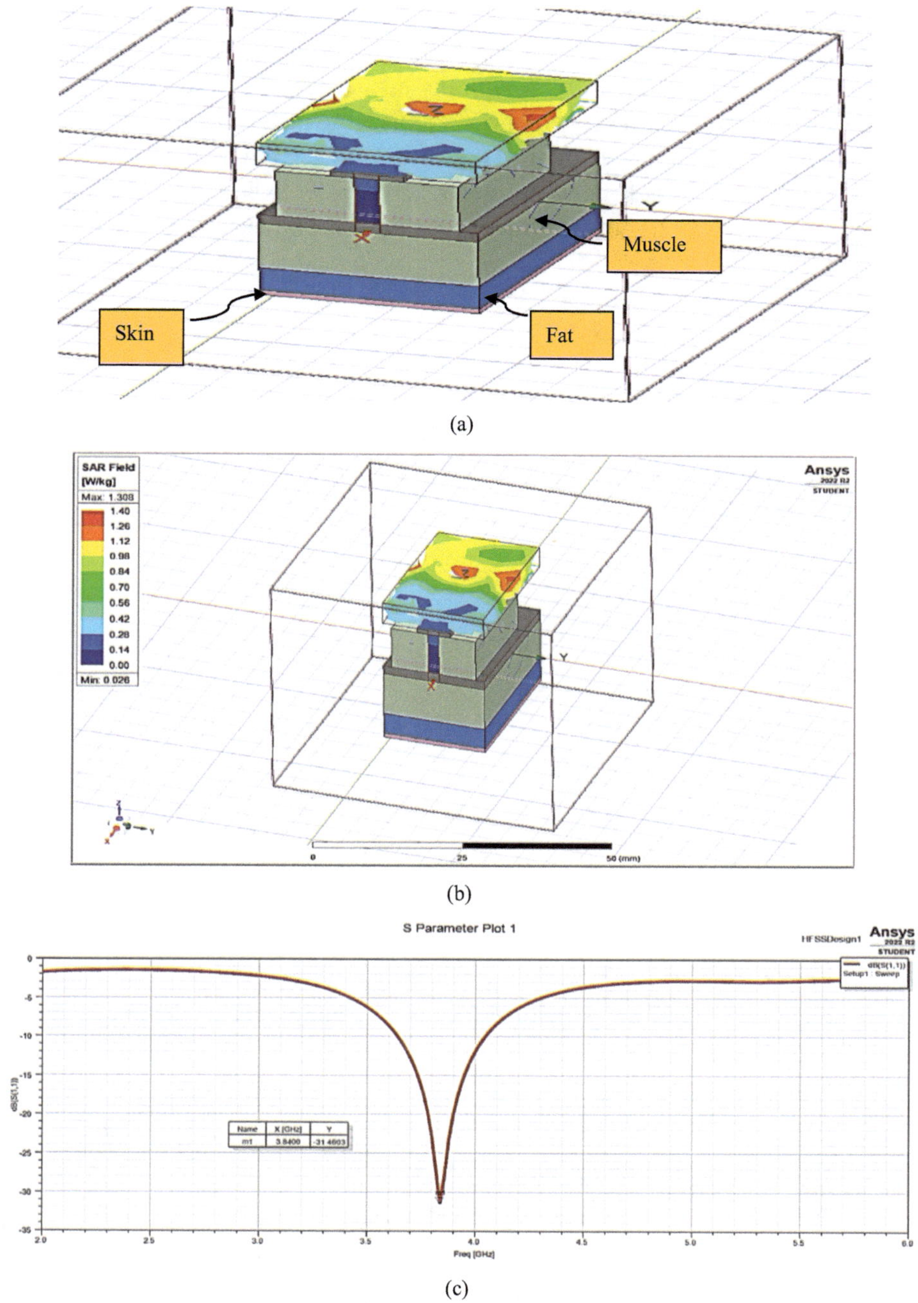

Fig. 7.8: Simulation of metastructure-based biosensor on the concept of MPA in HFSS (a), SAR calculation for the designed biosensor (b), and observed R_f (c).

step in the procedure involves using a roller to evenly distribute a thin layer of metal powder (usually 30–90 μm thick) throughout the build platform. According to the 2D digital slicing data, a powerful laser, typically a fiber laser, was focused onto the metal powder's surface. The energy of the laser raises the metal atoms in the targeted regions to the melting point, where they merge to fabricate the 3D model as per the CAD file. DMLS has several benefits, including the capacity to design complicated geometries, quick prototyping, and the manufacturing of useful metal components with superior mechanical qualities.

7.4.2 Lattice and infill pattern

For the fabrication of the patch, two types of lattice patterns were used as shown in Fig. 7.9. Two types of uniformly lattice cells are used: octet truss (Fig. 7.9(a)) and dodecahedron (Fig. 7.9(b)).

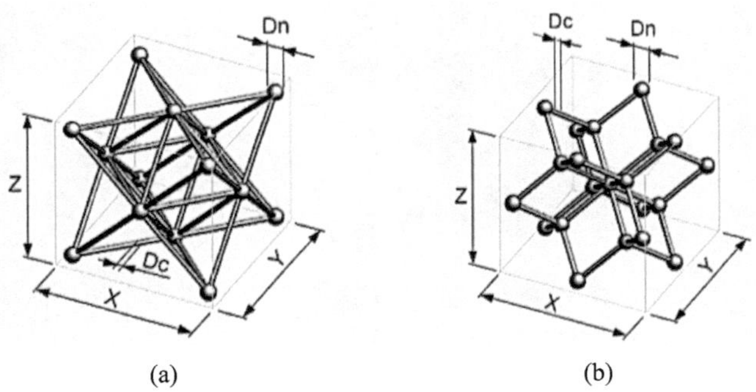

Fig. 7.9: Octet truss cell (a) and dodecahedron cell (b).

The parameters of the lattice structure (octet truss and dodecahedron) are given in Tab. 7.3.

Tab. 7.3: Parameters of octet truss cell and dodecahedron cell.

Parameter/cell	Octet truss cell	Dodecahedron cell
X	1.5	1.8
Y	1.6	2.3
Z	1.22	1.42
Dc	1.0	1.0
Dn	0.6	0.6

The selection of the sandwich-type metastructure design of the patch is shown in Fig. 7.10(a) Octet, and Fig. 7.10(b) Dodecahedron. The selection of metastructure for the fabrication of the patch is done from 3DXpert software.

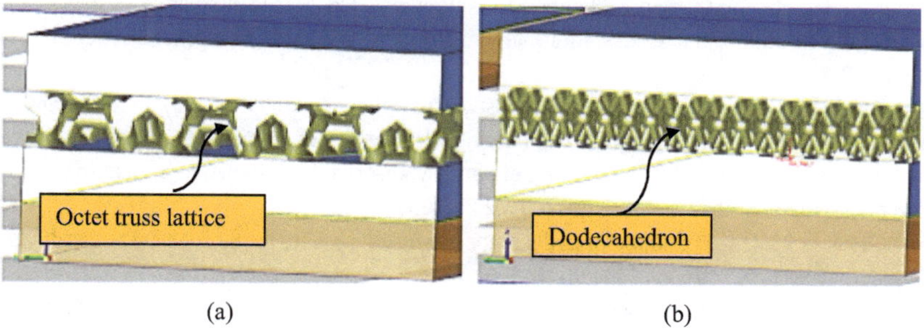

Fig. 7.10: (a) Octet truss lattice and (b) dodecahedron lattice.

The printing parameters used for the fabrication of the sandwich-type metastructure design of the patch and ground are given in Tab. 7.4.

Tab. 7.4: DMLS printing parameters.

S. no.	Input parameter	Value
1	Laser power (L_p)	130 W
2	Scanning speed (S_p)	600 mm/s
3	Hatch distance (H_d)	80 μm
4	Layer thickness (L_t)	30 μm
5	Energy density (E_d)	90.27 J/mm^3

7.5 Conclusion

This study highlights the investigation performed on different metastructures and the effect of these metastructures on resonance frequency (R_f) and specific absorption rate (SAR) value of biosensors. The fabrication of the patch and ground plane was printed by direct metal laser sintering (DMLS) process using two different metastructures (octet truss and dodecahedron) fixed in the patch and ground plane of the sensor for enhancing conformability. The result of this study suggests that the fabrication of a patch with octet truss metastructure has a better SAR value is 1.576 W/kg (without muscle, skin, and fat) and 1.40 W/kg with muscle, skin, and fat. The R_f and S_{11} without muscle, skin, and fat are 4.02 GHz and −16.25 dB, and with muscle, skin, and fat are 3.84 GHz and −31.46 dB, respectively for biosensors.

References

Brooks AK, Chakravarty S, Ali M, Yadavalli VK. Kirigami-inspired biodesign for applications in healthcare. Advanced Materials. 2022 May;34(18):2109550.

Dahri FA, Soomro RA, Memon SA, Memon Z, Memon MH. An optimized inset feed circular cross strip antenna design for c-band satellite links. International Journal of Advanced Computer Science and Applications. 2018;9(5).

Husain M, Kumar V, Singh G, Singh R. Meta structure-based 3D printing: A statistical analysis. Reference Module in Materials Science and Materials Engineering. Elsevier, 2023e. ISBN 9780128035818.

Husain M, Singh R, Pabla BS. A review on 3d printing of partially absorbable implants. Journal of the Institution of Engineers (India): Series C. 2023d Jul;23:1–20.

Husain M, Singh R, Pabla BS. On 3D printing of pvdf composite-based sensors for biomedical applications. National Academy Science Letters. 2023b Jun;22:1–6.

Husain M, Singh R, Pabla BS. On compressive properties of 3D printed polyvinylidene fluoride composite for implant applications. Journal of Thermoplastic Composite Materials. 2023c Sept;5:08927057231200005.

Husain M, Singh R, Pabla BS. On PVDF composite as partially absorbable smart implants. Proceedings of the Institution of Mechanical Engineers, Part H: Journal of Engineering in Medicine. 2023a Apr;237 (4):517–26.

Khaleghi A, Chávez-Santiago R, Balasingham I. Ultra-wideband pulse-based data communications for medical implants. IET Communications. 2010 Oct 15;4(15):1889–97.

McLaughlin RM. Surgical diseases of the feline stifle joint. Veterinary Clinics: Small Animal Practice. 2002 Jul 1;32(4):963–82.

Morgan JP, Wind A, Davidson AP. Hereditary bone and joint diseases in the dog: Osteochondroses, hip dysplasia, elbow dysplasia. Schlütersche. 2000.

Mostafa AA. Musculoskeletal Conformation of the Normal and Diseased Canine Stifle with Emphasis on Patella Luxation and Cranial Cruciate Ligament Deficiency. University of Illinois at Urbana-Champaign, 2008.

Mulcahy DM. Legal, ethical, and procedural bases for the use of aseptic techniques to implant electronic devices. Journal of Fish and Wildlife Management. 2013 Jun 1;4(1):211–19.

Nazir A, Abate KM, Kumar A, Jeng JY. A state-of-the-art review on types, design, optimization, and additive manufacturing of cellular structures. The International Journal of Advanced Manufacturing Technology. 2019 Oct;104:3489–510.

Nilsson K, Zanders S, Malm S. Heritability of patellar luxation in the Chihuahua and Bichon Frise breeds of dogs and effectiveness of a Swedish screening program. The Veterinary Journal. 2018 Apr 1;234:136–41.

Oshin A. Complications associated with the treatment of patellar luxation. Complications in Small Animal Surgery. 2016 Jan;8:889–96.

Pearce AI, Richards RG, Milz S, Schneider E, Pearce SG. Animal models for implant biomaterial research in bone: A review. European Cells & Materials. 2007 Mar 2;13(1):1–10.

Proffen BL, McElfresh M, Fleming BC, Murray MM. A comparative anatomical study of the human knee and six animal species. The Knee. 2012 Aug 1;19(4):493–99.

Rana MS, Rana SB, Rahman MM. Microstrip patch antennas for various applications: A review. Indonesian Journal of Electrical Engineering and Computer Science. 2023 Mar;29(3):1511–19.

Sarsamba MC, Yanamshetti R Microstrip antenna application for WiMAX, WLAN, mobile communication: A review. In2017 IEEE International Conference on Power, Control, Signals and Instrumentation Engineering (ICPCSI). 2017 Sep 21;988–91. IEEE.

Shahrubudin N, Lee TC, Ramlan RJ. An overview on 3D printing technology: Technological, materials, and applications. Procedia Manufacturing. 2019 Jan 1;35:1286–96.

Singh I, Tripathi VS. Micro strip patch antenna and its applications: A survey. International Journal of Computer Applications. 2011 Sep;2(5):1595–99.
Singh R, Husain M. Orthopedic implant fabrication for canine using DMLS: A case study. Reference Module in Materials Science and Materials Engineering. Elsevier, 2023. ISBN 9780128035818.
Stein S, Schmoekel H. Short-term and eight to 12 months results of a tibial tuberosity advancement as treatment of canine cranial cruciate ligament damage. Journal of Small Animal Practice. 2008 Aug; 49(8):398–404.
Ward SR, Terk MR, Powers CM. Patella alta: Association with patellofemoral alignment and changes in contact area during weight-bearing. JBJS. 2007 Aug 1;89(8):1749–55.
Yang D, Mei H, Yao L, Yang W, Yao Y, Cheng L, Zhang L, Dassios KG. 3D/4D printed tunable electrical metamaterials with more sophisticated structures. Journal of Materials Chemistry C. 2021;9 (36):12010–36.

Gurwinder Singh, Rupinder Singh, Minhaz Husain*, Vinay Kumar, J. Paulo Davim

8 Partially absorbable 3D-printed implant for health monitoring

Abstract: In the past decade, many researchers have reported on the online health monitoring of animals and humans using various techniques like Zigbee module, mobile ad network, zero-time effort technology, and microstrip patch antenna. However little has been reported on the partially absorbable 3D-printed implant sensor and the effect of the substrate's thickness on the sensor's performance for online health monitoring of animals and humans. This study details partially absorbable 3D-printed implant sensors and the role of substrate thickness for online health monitoring of animals. An application of sensors in cattle (buffalo) is also highlighted, which helps to monitor the health conditions of bovine. This implantable sensor was composed of polyvinylidene fluoride composite, polylactic acid, and 17–4 precipitation hardened stainless steel (SS).

***Corresponding author: Minhaz Husain,** Mechanical Engineering Department, National Institute of Technical Teachers Training and Research, Chandigarh, India, e-mail: minhazme2012@gmail.com
Gurwinder Singh, Rupinder Singh, Mechanical Engineering Department, National Institute of Technical Teachers Training and Research, Chandigarh, India
Vinay Kumar, University Centre of Research and Development, Chandigarh University, Mohali, Punjab, India
J. Paulo Davim, Mechanical Engineering Department, University of Aveiro, Portugal

https://doi.org/10.1515/9783111081632-008

Graphical abstract:

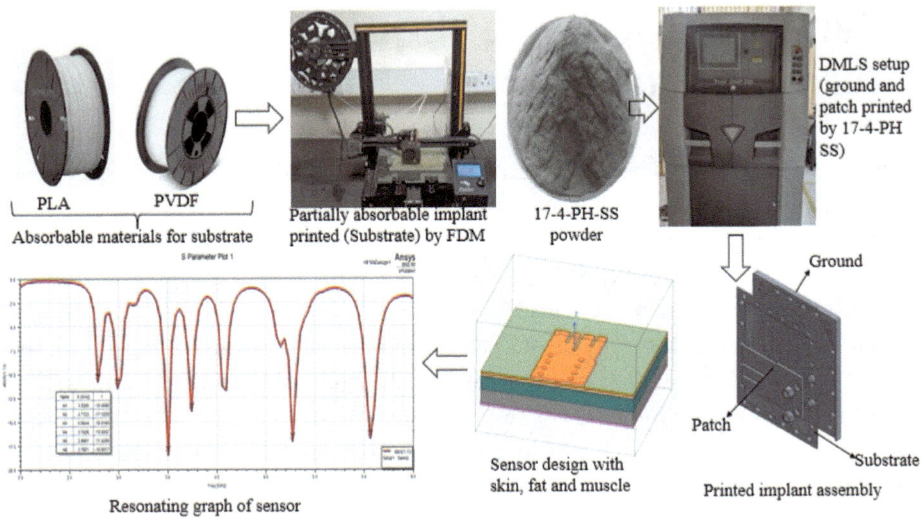

Keywords: 3D printing, health monitoring, microstrip patch antenna, substrate, diaphragmatic hernia

8.1 Introduction

The additive manufacturing (AM) process adopts a 3D printing philosophy in which functional/nonfunctional parts are assembled layer by layer. These techniques/processes are employed not only to create different mechanical parts but also to fabricate electrical devices and sensors (Khosravani and Reinicke 2020). Antennas, capacitors, resistors, and inductors can all possibly be manufactured by AM methods in a single production step (removing the need for any additional processing) (Lee et al. 2017). Various researchers have employed 3D printing (Park et al. 2019) in the realm of biomedical applications. The 3D bioprinting technique is being utilized to produce a wireless pressure sensor and a smart stent crafted from polymer materials. Park et al. (2015) fabricated a drug-coated bioabsorbable vascular polycaprolactone stent using an ultrasonic spray coating system. Chang et al. (2020) fabricated a 3D-printed radiopaque polymer composite (PGSA-BiOCl) and monitored the degradation of material for biomedical purposes. Rifai et al. (2021) fabricated a 3D-printed diamond for orthopedic applications using the selective laser melting technique. Commercially, microstrip patch antennas (MPAs) are fabricated by placing a conductor on a substrate backed by a ground plane. MPA, particularly in circular and rectangle patch shapes, are commonly employed due to their straightforward and well-established design process. These antennas offer several advantages including ease of fabrication and integration with other electronics components, cost-effectiveness due to the ability to be produced on various affordable substrates, durability, and a low-profile design. These qualities make MPA highly suitable for numerous practical applications (Khan et al. 2015). Singh and Tripathi (2011) outlined four types of feeding techniques used in MPA, that is, coaxial probe feed, microstrip line feed, aperture couple feed, and proximity coupling feed. MPA has been used in varied military and commercial applications such as radar, satellite, biomedical, automotive, mobile communication, wireless local area networks, ultra-wideband systems, and for radio frequency identification (Peixeiro 2011). Panda et al. (2021) proposed a wearable antenna for health and weight monitoring using different substrate materials. Amit et al. (2020) proposed a wearable antenna for monitoring health parameters like body temperature and heart rate. Ariff et al. (2014) proposed an ultra-high-frequency (UHF) radio frequency identification antenna for health monitoring of live stocks, and an implantable antenna for medical telemetry in the human body. Recently, numerous biomedical applications to discern brain cancer (Chakraborty et al. 2018) and breast cancer (Raihan et al. 2017, Nesasudha 2021, Fusic et al. 2023) have been reported by using this technique. It is also used to measure the dielectric property of porcine skin tissue using a pig as a model for the implantable purpose (Karacolak et al. 2012) and to evaluate the psychological and physiological well-being of dairy cattle (Bjarnason et al. 2013). Deshmukh and Chorage (2020) highlighted the determination of glucose blood levels in the in-body patch antenna. It has been reported that in the industrial, scientific, and medical band, cardiac functions like heart rate may also be measured (Hasan et al. 2019). Chellathurai et al. (2023) proposed an implantable antenna for measuring ECG of the human body.

8.1.1 Role of substrate in MPA

a. Resonant frequency: The MPA frequency is influenced by the substrate's height. The frequency decreases with an increase in the substrate's height. This is because variations in the resonant frequency result from variations in the effective electrical length of the patch, which is influenced by the height of the substrate (Balanis 2016).
b. Bandwidth: The antenna's bandwidth is influenced by the substrate height. A broader (larger) bandwidth is often provided by a thicker substrate whereas a narrower (smaller) bandwidth is produced by the thinner substrate. This is due to the influence that the substrate height has on the microstrip's effective dielectric constant, which affects the phase velocity and impedance bandwidth.
c. Efficiency: The antenna's overall efficiency may be impacted by the substrate height. By reducing surface wave losses and increasing antenna gain, increasing the substrate height in specific situations can increase efficiency. However, a thicker substrate may also result in increased dielectric losses, which might lead to a reduction in efficiency.

Various researchers employed a 3D-printed technique for the fabrication of the substrate. Singh et al. (2022) used a substrate of recycled low-density polyethylene (LDPE) and LDPE-Bakelite through the fused filament fabrication technique. Samal et al. (2012) used a silicon substrate for implantable purposes. Feili et al. (2010) used liquid crystal polymer (LCP) substrate for flexible wireless applications like aircraft monitoring, military devices, etc. Prakash et al. (2021) used a polylactic acid (PLA) substrate printed by fused deposition modeling (FDM). Tesser et al. (2019) used an antenna of PLA-sunflower carbon substrate as a substrate for the health monitoring of humans. Husain et al. (2023) used a polyvinylidene fluoride (PVDF) composite printed by FDM for implant purposes. Lee et al. (2020) used hexaferrite particles ($Ba_{1.5}Sr_{1.5}Z$). In particular, they were incorporated into polymer composites, with polyurethane serving as the substrate for antennas used in the medical industry that operates in the very high frequency and low UHF bands. These composites are crucial parts of implantable medical devices. Farooq et al. (2019) proposed a PET substrate for wearable and medical purposes. From the literature, it has been concluded that many researchers have used different techniques for online health monitoring of animals as well as have used different substrate materials, but little has been reported on the partially absorbable 3D-printed implant for health monitoring. So, this paper highlights the partially absorbable 3D-printed implant for health monitoring, along with its application.

8.2 Research gap

For the research gap, the database of Web of Science of the past 20 years was explored using the keywords "substrate for online health monitoring" between 2003 and 2023 and selection was processed on VOS viewer software for bibliographic analysis. Relevance scores are given in Tab. 8.1 with selected keywords. Figure 8.1 shows a networking diagram of the dielectric substrate for online health monitoring. Figure 8.2 shows the networking of the antenna with substrate thickness, and Fig. 8.3 shows the networking of dielectric property for health monitoring. Finally, based on Fig. 8.4, the research gap has been defined for this proposed work.

Tab. 8.1: Number of occurrences.

ID	Terms	Occurrences	Relevance score
1	ABS	5	1.2163
2	Active layer	6	1.7838
3	Antenna	68	0.3552
4	Antenna sensor	14	0.3836
5	Band	14	0.557

Tab. 8.1 (continued)

ID	Terms	Occurrences	Relevance score
6	Capacitance	12	2.2868
7	Capacitive pressure sensor	11	1.6027
8	Capacitive sensor	7	3.3301
9	Cell	31	1.4776
10	Cu coating sensor	5	5.5087
11	DBI	5	1.0512
12	Deformation	9	0.5002
13	Dielectric layer	13	2.1345
14	Dielectric permittivity	5	0.5546
15	Dielectric property	12	0.4917
16	Dielectric substrate	15	0.1065
17	Electromagnetic field	5	0.4763
18	Electronic skin	6	1.2856
19	Eye	7	0.6015
20	Flexible substrate	11	0.9683
21	Gain	11	0.6621
22	GHz	22	0.4773
23	Graphene	16	1.0891
24	Ground plane	5	0.3242
25	Healthcare	5	0.8687
26	Internet	6	0.3503
27	Low cost	14	0.2237
28	Magnitude	6	0.429
29	Medical application	6	0.4296
30	Metal	8	0.7585
31	Model	28	0.3221
32	Module	5	0.571
33	Patch	21	0.3529
34	Plastic substrate	5	0.3857
35	Polymer	7	1.2643
36	Presence	9	1.6178
37	Pressure	12	1.1992
38	Pressure sensor	14	1.7992
39	Real time	6	1.8656
40	Resonant frequency	12	0.1793
41	SAR	6	0.8483
42	Strain	24	0.3172
43	Strain sensor	8	0.432
44	Temperature	29	0.3336
45	Thickness	12	0.4646
46	Thin film	18	1.1263
47	Treatment	7	1.9659
48	Vector network analyzer	7	0.6856
49	VNA	5	1.0341
50	Wearable antenna	6	0.951

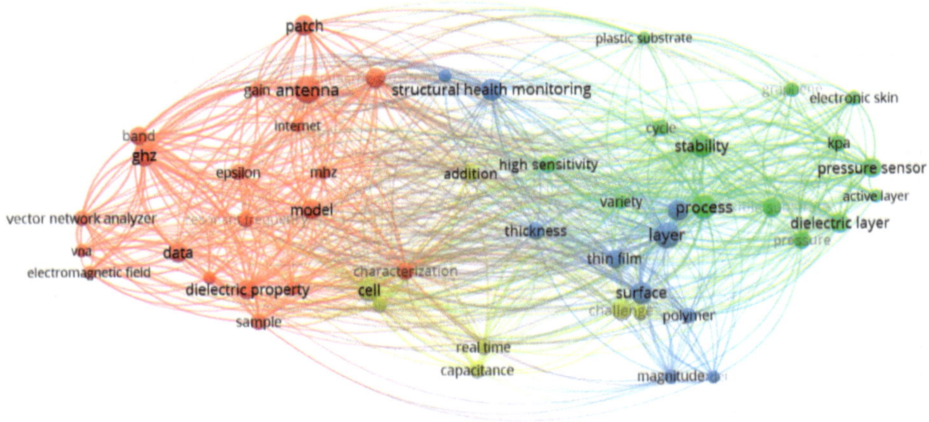

Fig. 8.1: Network diagram of "dielectric substrate for online health monitoring."

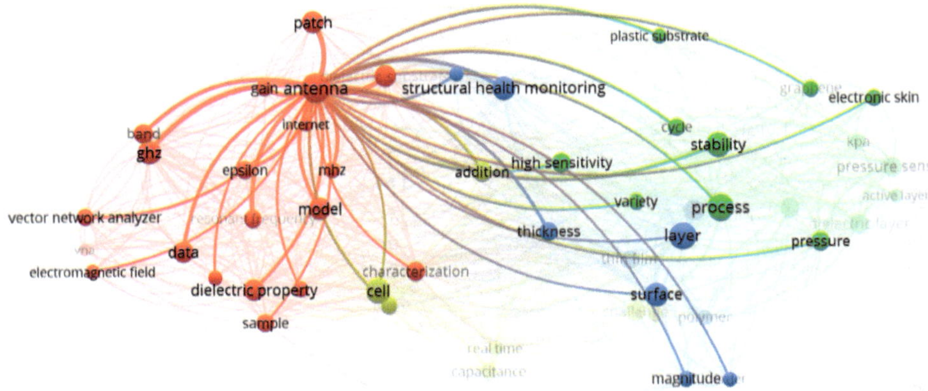

Fig. 8.2: Analysis of the keyword "antenna with substrate thickness."

8.3 Case study

In milk buffaloes, diaphragmatic hernia is a prevalent thoraco-abdominal disease. Due to the severe morbidity and mortality associated with this disease, it has become one of the leading surgical conditions influencing farm animal production. The primary causes of reticular diaphragmatic hernia in buffaloes include increased intra-abdominal pressure brought on by the viscera's usual heavyweight, variable degrees of tympany, advanced pregnancy, and efforts at calving (Pal and Tayal 2018). Reticular diaphragmatic hernia in cows and buffaloes has been recorded as both a congenital and acquired disorder (Kumar and Saini 2011). Some symptoms of DH include dyspnea, recurrent tympany,

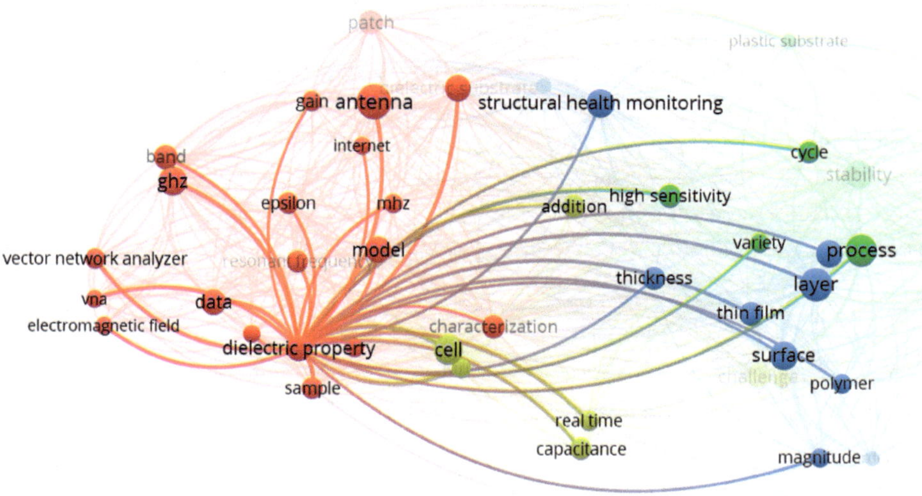

Fig. 8.3: Analysis of the keyword "dielectric property for health monitoring."

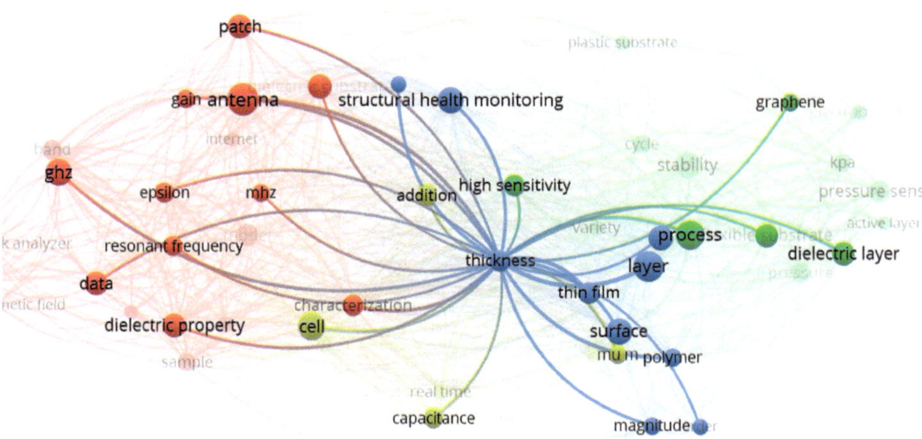

Fig. 8.4: Research gap.

partial-to-full anorexia, regurgitated items coming out from the mouth and nose, and suspended rumination (Talekar et al. 2018). Due to the absence of pericardiacophrenic arteries and Golgi tendon organs, which serve as pressure sensors in the diaphragm, buffaloes are more susceptible to diaphragmatic hernia than cattle. Compared to heifers and nonpregnant buffaloes, pregnant buffaloes experience DH more frequently (Pooniya et al. 2022). Conventional approaches for diagnosing DH include clinical symptoms, radiography, ferroscopy, ultrasonography, and hematobiochemical tests; however, the rate of success is insufficient (Bhimappa et al. 2023). Therefore, a creative IOT-based approach for tracking the recurrence of DH in bovine buffalo has been employed.

Figure 8.5(a, b) shows a sensor based on the anatomy of the buffalo stomach. PVDF polymer and 17-4 PH (precipitation hardened) stainless steel (SS) passive sensors were used in this study (Indian patent application no.: 202311036753). After surgery on a bovine, the implantable sensor may help in tracking the recurrence of diaphragmatic hernia. The subparts of the sensor's assemblies were produced using an open-source 3D polymer printer and direct metal laser sintering (DMLS) setup.

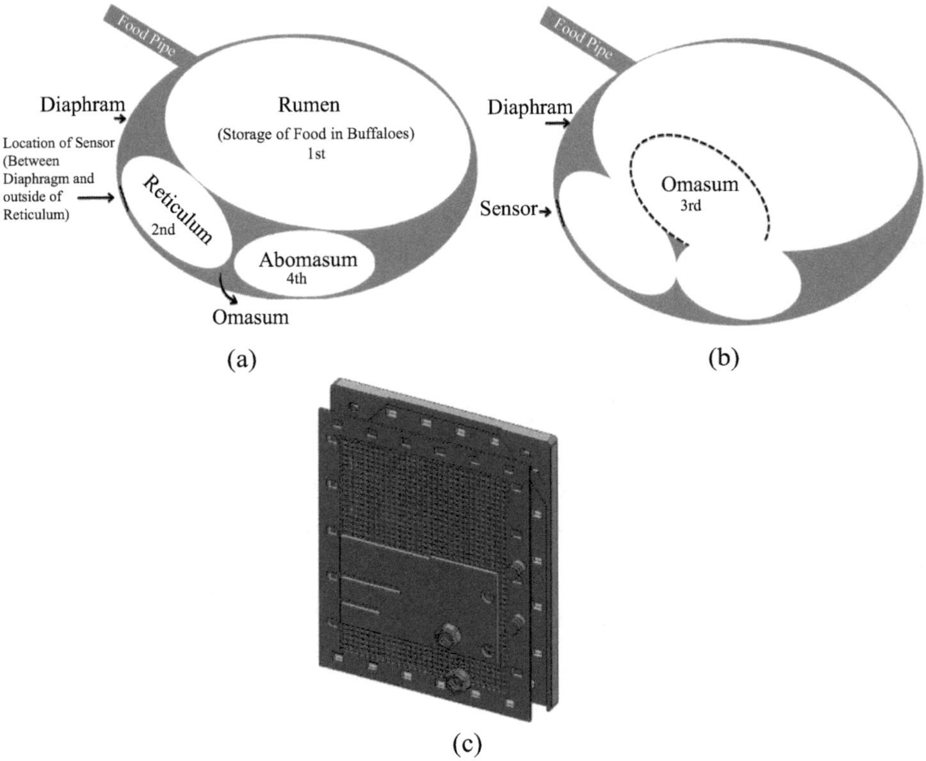

Fig. 8.5: Anatomy of buffalo stomach (a) looking from the right side and (b) left side, and (c) assembly view of the sensor.

As per the concept of the MPA, the patch and the ground were made of DMLS-fabricated 17-4 PH SS, while the mesh was made of PVDF and the hexagonal bolt head was made of PLA. While doing surgery, the sensor was attached to the diaphragm outside the reticulum. The PVDF substrate shifts from alpha to beta phase and ultimately alters the material's dielectric constant, and this signal can be picked up by an electronic device when some metallic or nonmetallic foreign substances press against the sensor.

8.4 Results and discussion

Three cases are shown for health monitoring of cattle using MPA. It has been observed that with a change in substrate thickness, there is a change in resonating frequency.

Case 1: For a 1-mm thick substrate, the sensor design is shown in Fig. 8.6 and the graph for reflection coefficient (S_{11}) and resonating frequency is shown in Fig. 8.7.

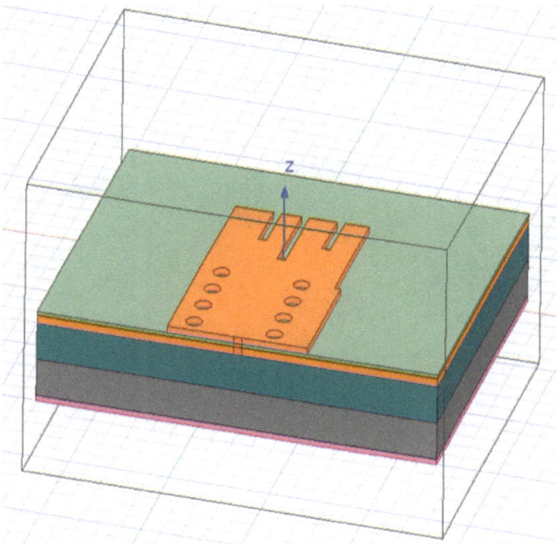

Fig. 8.6: Sensor design for a substrate of 1 mm.

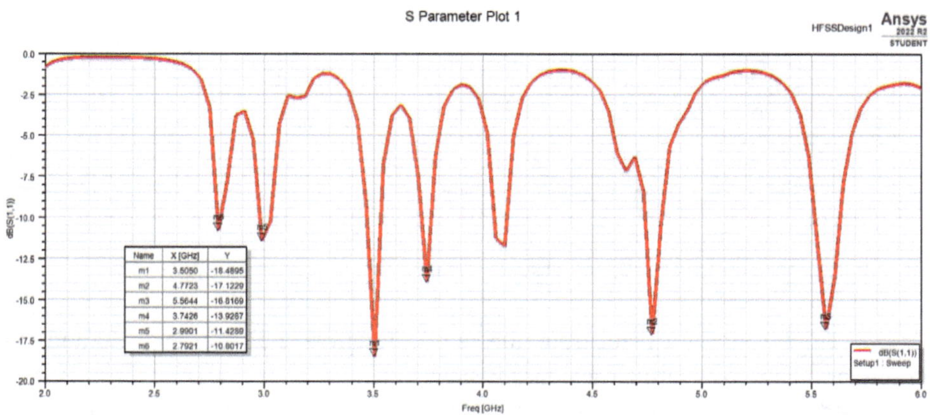

Fig. 8.7: Resonating graph versus S_{11} for substrate of 1 mm.

Case 2: For a 2-mm thick substrate, the sensor design is shown in Fig. 8.8 and the graph for S_{11} and resonating frequency is shown in Fig. 8.9.

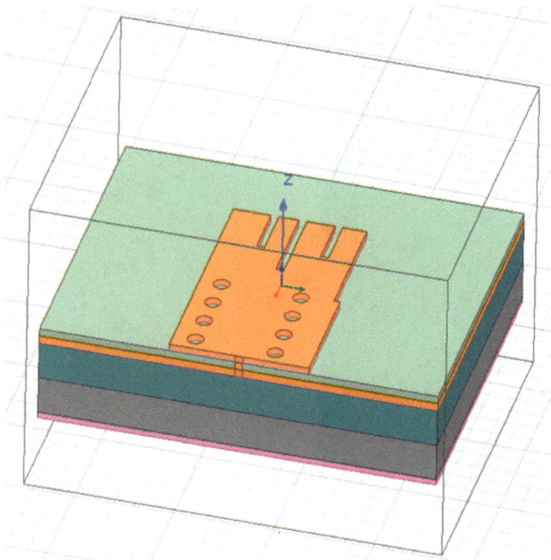

Fig. 8.8: Sensor design for substrate of 2 mm.

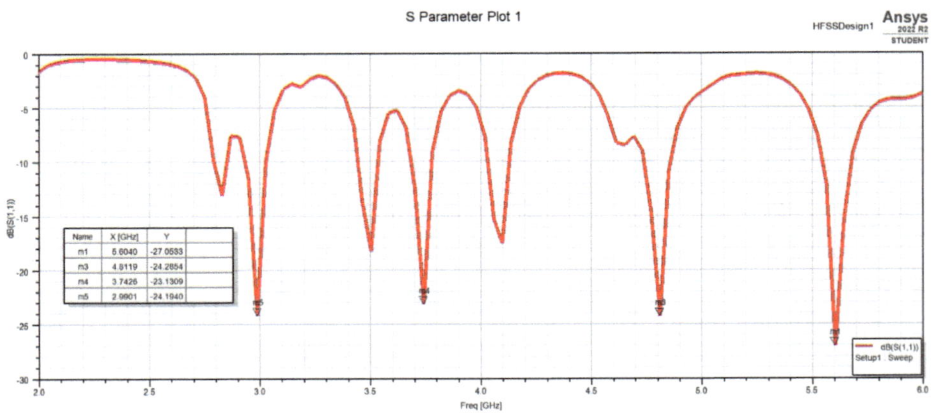

Fig. 8.9: Resonating graph versus S_{11} for substrate of 2 mm.

Case 3: For a 3-mm thick substrate, the sensor design is shown in Fig. 8.10 and the graph for S_{11} and resonating frequency is shown in Fig. 8.11.

Based on Figs. 8.6–8.11, it has been noticed that the change in substrate thickness has led to a change in resonating frequency as well as in the S_{11} parameters, which may be tuned to assess the health of the patient.

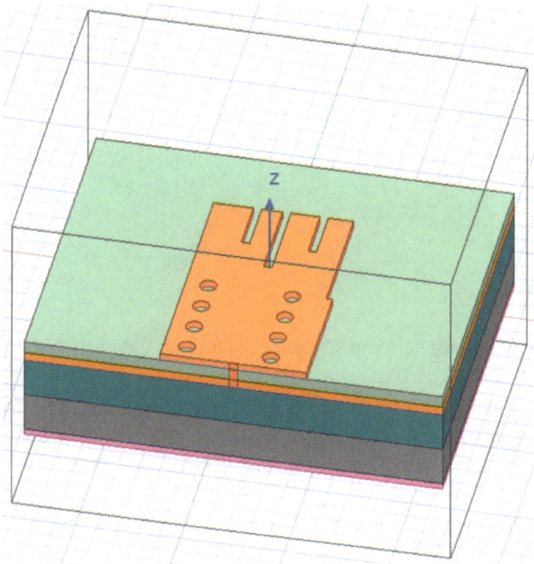

Fig. 8.10: Sensor design for a substrate of 3 mm.

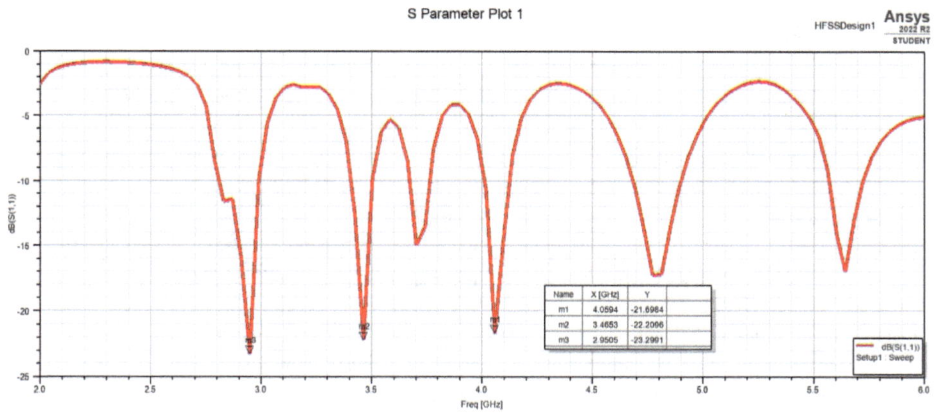

Fig. 8.11: Resonating graph versus S_{11} for substrate of 3 mm.

8.5 Conclusions

In this work, a 3D-printed partially absorbable implant for online health monitoring has been investigated. The following are the conclusions from this study:

It has been noticed that the change in substrate thickness has led to a change in resonating frequency as well as in S_{11} parameters, which may be tuned to assess the health of the patient. Finite element analysis may be conducted in future studies to ensure the effect of cycling loading on the sensor. Further studies may be performed for different biocompatible piezoelectric substrate materials, along with different conducting materials. In vitro studies need to be conducted before actual implantation into the patient's body. Also, specific absorption ratio values need to be computed to ensure the sensor is working in an acceptable range.

References

A post-surgery detecting device for diaphragmatic hernia in bovines, Application filed on Dated: 27/05/2023, Indian Patent Application No.: 202311036753.

Amit S, Talasila V, Nagaveni CR, Subbaraya CK, Mahesh DS, Premnath PM. Design and development of wearable antenna integrated with RF transceiver for health monitoring system. In2020 International Conference on Recent Trends on Electronics, Information, Communication & Technology (RTEICT). 2020 Nov;12:242–46. IEEE.

Ariff MH, Ismarani I, Shamsuddin N. Design and development of UHF RFID reader antenna for livestock monitoring. In2014 IEEE 5th Control and System Graduate Research Colloquium. 2014 Aug;11:125–29. IEEE.

Balanis CA. Antenna Theory: Analysis and Design. John Wiley & sons, 2016 Feb 1.

Bhimappa A, Dilipkumar D, Bhagavantappa B, Manjunatha DR, Patil M, Halmandge S, Suranagi MD. Radiographic and ultrasonographic diagnosis as well as surgical treatment of foreign body-induced reticular affections in bovines.

Bjarnason A, Vuorela T, Verho J, Riistama J, Vanhala J, Lekkala J, Hyttinen J. Implantable measurement system for dairy-cattle monitoring with long recording time. Advances in Science and Technology. 2013 Jan 13;85:33–38.

Chakraborty S, Islam AJ, Farhad MM, Hasan MM, Nesar MS, Rahaman MA. Performance studies of UWB microstrip antenna for multipurpose biotelemetry applications. In2018 International Conference on Innovations in Science, Engineering and Technology (ICISET). 2018 Oct;27:322–27. IEEE.

Chang CT, Chen HT, Girsang SP, Chen YM, Wan D, Shen SH, Wang J. 3D-printed radiopaque polymer composites for the in situ monitoring of biodegradable medical implants. Applied Materials Today. 2020 Sep 1;20:100771.

Chellathurai AS, Sugumar D, Kishor T, Jacob LS, Aldrin AA, Shamith M. Performance analysis of novel coplanar waveguide fed slot antenna for ECG monitoring. In2023 4th International Conference on Signal Processing and Communication (ICSPC). 2023 Mar;23:334–38. IEEE.

Deshmukh VV, Chorage SS. Microstrip antennas used for noninvasive determination of blood glucose level. In2020 4th International Conference on Intelligent Computing and Control Systems (ICICCS). 2020 May;13:720–25. IEEE.

Farooq U, Iftikhar A, Fida A, Khan MS, Shafique MF, Asif SM, Shubair RM. Design of a 1× 4 CPW microstrip antenna array on PET substrate for biomedical applications. In2019 IEEE International Symposium on Antennas and Propagation and USNC-URSI Radio Science Meeting. 2019 Jul;7:1345–46. IEEE.

Feili D, Pagel N, Ngo VL, Schwarz P, Seidel H. Cu-based patch antenna on polymer substrate for flexible wireless sensor systems applications. Procedia Engineering. 2010 Jan 1;5:890–93.

Fusic J, Sugumari T, Sitharthan R. Design of a wearable, flexible Microstrip patch antenna for the detection of breast cancer.

Hasan RR, Rahman MA, Sinha S, Uddin MN, Niloy TS. In-body antenna for monitoring pacemaker. In2019 International Conference on Automation, Computational and Technology Management (ICACTM). 2019 Apr;24:99–102. IEEE.

Husain M, Singh R, Pabla BS. On PVDF composite as partially absorbable smart implants. Proceedings of the Institution of Mechanical Engineers, Part H: Journal of Engineering in Medicine. 2023 Apr;237 (4):517–26.

Karacolak T, Cooper R, Unlu ES, Topsakal E. Dielectric properties of porcine skin tissue and in vivo testing of implantable antennas using pigs as model animals. IEEE Antennas and Wireless Propagation Letters. 2012;11:1686–89.

Khan MU, Sharawi MS, Mittra R. Microstrip patch antenna miniaturisation techniques: A review. IET Microwaves, Antennas & Propagation. 2015 Jun;9(9):913–22.

Khosravani MR, Reinicke T. 3D-printed sensors: Current progress and future challenges. Sensors and Actuators A: Physical. 2020 Apr 15;305:111916.

Kumar A, Saini NS. Reliability of ultrasonography at the fifth intercostal space in the diagnosis of reticular diaphragmatic hernia. Veterinary Record-English Edition. 2011 Oct 8;169(15):391.

Lee JY, An J, Chua CK. Fundamentals and applications of 3D printing for novel materials. Applied Materials Today. 2017 Jun 1;7:120–33.

Lee SE, Choi SP, Oh KS, Kim J, Lee SM, Cho KR. Flexible magnetic polymer composite substrate with Ba1. 5Sr1. 5Z Hexaferrite Particles of VHF/Low UHF Patch Antennas for UAVs and medical implant devices. Materials. 2020 Feb 24;13(4):1021.

Nesasudha M. Microstrip antenna for early-stage breast cancer detection – A survey. Health and Technology. 2021 Nov;11:1191–204.

Pal M, Tayal R. Diagnosis of diaphragmatic hernia in buffaloes: A review. Indian Journal of Veterinary Sciences & Biotechnology. 2018 Apr 8;13(4):78–83.

Panda S, Gupta A, Acharya B. Wearable microstrip patch antennas with different flexible substrates for health monitoring system. Materials Today: Proceedings. 2021 Jan 1;45:4002–07.

Park J, Kim JK, Kim DS, Shanmugasundaram A, Park SA, Kang S, Kim SH, Jeong MH, Lee DW. Wireless pressure sensor integrated with a 3D printed polymer stent for smart health monitoring. Sensors and Actuators B: Chemical. 2019 Feb 1;280:201–09.

Park SA, Lee SJ, Lim KS, Bae IH, Lee JH, Kim WD, Jeong MH, Park JK. In vivo evaluation and characterization of a bio-absorbable drug-coated stent fabricated using a 3D-printing system. Materials Letters. 2015 Feb 15;141:355–58.

Peixeiro C. Microstrip patch antennas: An historical perspective of the development. In2011 SBMO/IEEE MTT-S International Microwave and Optoelectronics Conference (IMOC 2011). 2011 Oct;29:684–88. IEEE.

Pooniya R, Palsania SK, Saini R, Kumar A, Lal M, Kumar H. Surgico-therapeutic management of diaphragmatic hernia in buffalo: A review of 8 cases.

Prakash C, Senthil P, Sathies T. Fused deposition modeling fabricated PLA dielectric substrate for microstrip patch antenna. Materials Today: Proceedings. 2021 Jan 1;39:533–37.

Raihan R, Bhuiyan MS, Hasan RR, Chowdhury T, Farhin R. A wearable microstrip patch antenna for detecting brain cancer. In2017 IEEE 2nd International Conference on Signal and Image Processing (ICSIP). 2017 Aug 4;432–36. IEEE.

Rifai A, Houshyar S, Fox K. Progress towards 3D-printing diamond for medical implants: A review. Annals of 3D Printed Medicine. 2021 Mar 1;1:100002.

Samal S, Dwari S, Dutta A, Reddy SP. A microstrip patch antenna for biomedical applications at 2.45 GHz. In2012 5th International Conference on Computers and Devices for Communication (CODEC). 2012 Dec;17:1–4. IEEE.

Singh I, Tripathi VS. Micro strip patch antenna and its applications: A survey. International Journal of Computer Technology and Applications. 2011 Sep;2(5):1595–99.

Singh R, Kumar S, Singh AP, Wei Y. On comparison of recycled LDPE and LDPE–bakelite composite based 3D printed patch antenna. Proceedings of the Institution of Mechanical Engineers, Part L: Journal of Materials: Design and Applications. 2022 Apr;236(4):842–56.

Talekar SH, Chaudhary MP, Gamiti KS, Katara KP, Manish S. Surgical management of diaphragmatic hernia in four Jaffarabadi buffaloes and two Gir cows. International Journal of Science, Environment and Technology. 2018;7(2):684–88.

Tesser A, Alemaryeen A, Lewis J, Alshami A, Haghshenas M, Noghanian S. Synthesis and use of bio-based dielectric substrate for implanted radio frequency antennas. IEEE Access. 2019 Aug 27;7:123268–79.

Ankush Mehta*, Rupinder Singh, Bahadur Singh Pabla

9 Smart foot sensors by 3D bioprinting

Abstract: The pressure measurement of the foot plays a significant role in numerous applications such as sports, medical rehabilitation, and healthcare. In the past few decades, numerous pressure-sensing techniques have been developed for such cases. But hitherto, little has been reported on the 3D-printed smart insole foot sensors. This study highlights the sensing capability in terms of dielectric constant (ε_r) of 3D-printed smart foot sensors for continuous health monitoring of the human feet. The continuous health monitoring of the foot through this smart insole foot sensor may help determine the pressure distribution during normal walking and running for a healthy and unhealthy person. The study also highlights the detailed procedure for the calculation of ε_r and results have been analyzed and compared through experimentation and simulation.

***Corresponding author: Ankush Mehta,** Department of Mechanical Engineering, NITTTR, Chandigarh, India, e-mail: ankush.mehta38@gmail.com
Rupinder Singh, Bahadur Singh Pabla, Department of Mechanical Engineering, NITTTR, Chandigarh, India

https://doi.org/10.1515/9783111081632-009

Graphical abstract:

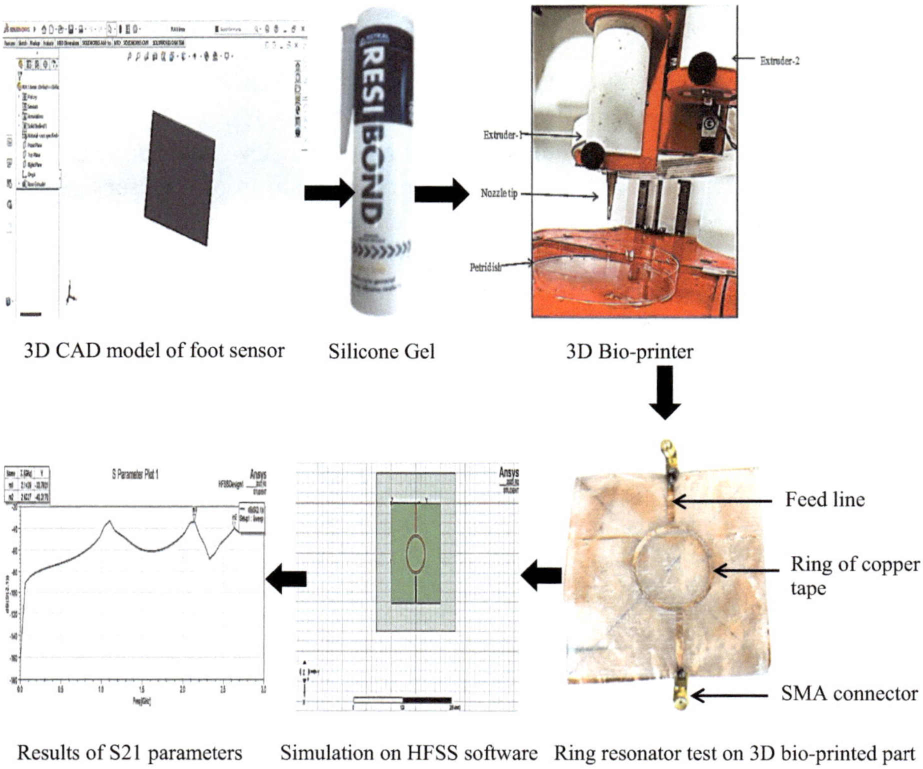

Keywords: 3D bioprinting, smart foot sensor, dielectric constant, ring resonator, health monitoring

9.1 Introduction

Insole foot sensors have become an important aspect in the fields of sports and healthcare as improper pressure distribution over the foot causes problems such as diabetes foot ulcers. It has been reported by the International Diabetes Federation that every year almost 1.1 million people in Europe suffer from diabetic foot ulcers (Federation et al. 2019, Husain et al. 2023a, b). Any abrupt changes in unbalanced gait are the primary indicators of diabetic foot ulcer, the effect of which can be reduced to 85% if diagnosed at the initial stage (Chatwin et al. 2020, Barkley et al. 2011). The condition monitoring of gait plays a vital role during sports training since an unbalanced gait directly affects the psychology of a sportsperson and can be responsible for sports-associated injuries (Prompers et al. 2008). It was also reported by the researchers that poor posture and unbalanced gait may lead to temporomandibular disorder and to get rid of this disorder, researchers have reported the influence of therapeutic exercise and occlusal splint on the posture and temporomandibular disorder (Oliveira et al. 2019, Moccia et al. 2020). Therefore, there is an urgent need to monitor the distribution of pressure between the ground and foot plantar surface to avoid sports-related injuries and the monitoring of pressure helps in identifying and rehabilitation foot injuries (Avedesian et al. 2020). Studies have also been reported on low-cost pressure sensing techniques such as optical fibers (Leal-Junior et al. 2020) and air pressure with an air bladder (Yang et al. 2020). The maximum range of these techniques is up to 150kPa; however, these are mostly suitable for static load and lose their durability in the dynamic mode (Razak et al. 2012, Husain et al. 2023c). Studies have also reported on electrical sensors placed in shoes for the measurement of foot pressure. These in-shoe pressure sensors are flexible and measure the pressure between the shoe and the foot and studies have reported on various commercially available foot pressure sensing equipment such as Orpyx LogR, Pedar, Medilogic, and Tekscan, which are available in the open market (Razak et al. 2012, Jor et al. 2019, Cicirelli et al. 2021, Rajendran et al. 2021, Minty et al. 2023). These sensing devices are not inexpensive; hence, to overcome this limitation, various studies have been reported on low-cost piezoresistive, piezoelectric, and capacitive pressure sensors (Ramirez et al. 2017). It has been reported that polyvinylidene fluoride (PVDF) piezoelectric sensor has been used both to measure the pressure as well as for energy harvesting and almost 1.6 pJ energy was stored by the PVDF insole sensor for a complete walking cycle (Chen et al. 2021, Husain et al. 2023a, d). Another foot sensor as a sandwich structure consisting of fluorinated ethylene propylene and cotton was used as a ferroelectric textile–based foot sensor for pressure measurement (Shi et al. 2019). Studies have also reported on polymers such as PVDF, lead zirconate titanate (PZT), etc., for the fabrication of foot sensors due to their higher impedance and lower signal-to-noise ratio (Lekkala et al. 2014). Studies have also reported on capacitive foot sensors due to their high linearity and sensitivity for the measurement of pressure (Aqueveque et al. 2018, Tao et al. 2020, Jung et al. 2020). Studies have reported on the monitoring of pressure distribution by using a smart insole 3D-printed capacitive foot sensor as shown in Fig. 9.1 (Samarentsis et al. 2022).

9.2 Background

To study the gaps in the literature, bibliographic analysis was performed. For bibliographic analysis, Web of Science data of the past 20 years for the keyword "foot sensor and biomedical applications" has been used and their network diagram is shown in Fig. 9.2.

The Web of Science database for the selected keywords foot sensor and biomedical applications gives the results, which were sorted based on the maximum citations and the top 34 articles were chosen. The data of these selected articles were entered

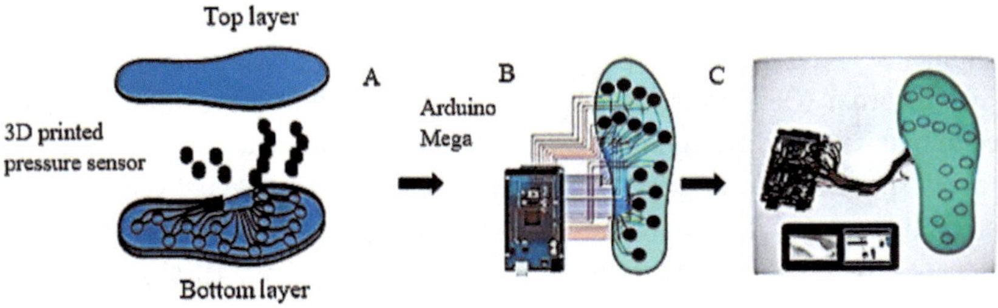

Fig. 9.1: Schematic diagram of 3D-printed insole (A), circuit diagram of foot sensor using Arduino (B), and smart 3D-printed capacitive smart insole foot sensor (C).

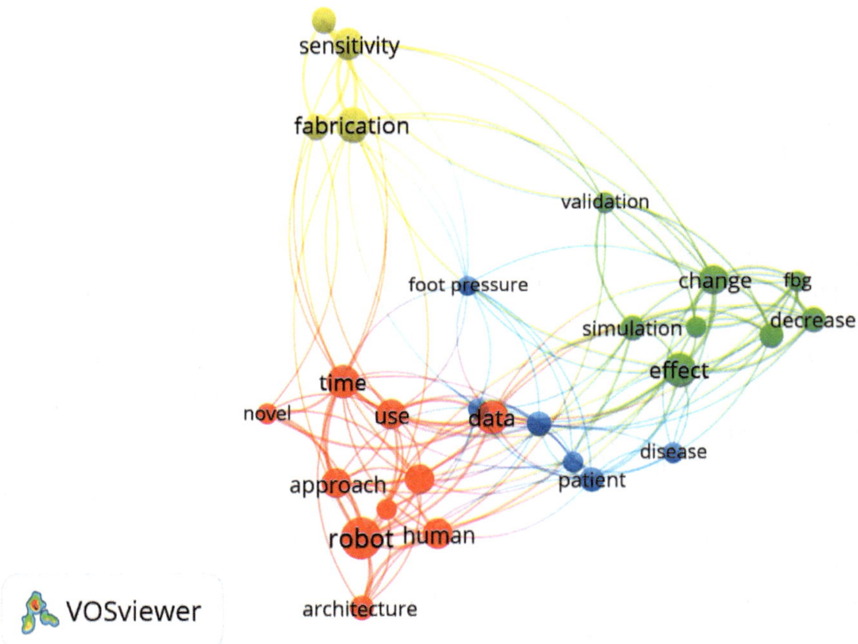

Fig. 9.2: Bibliographic analysis based on keyword "foot sensor and biomedical applications."

into VOSviewer software. The minimum occurrence of the term is 3; out of 1,421, 47 terms meet the threshold; a relevance score for the 47 terms was considered and 60% of terms (28) were used for the analysis. The number of terms, occurrences, and the relevance scores are given in Tab. 9.1.

The literature of the past 20 years has been reviewed and it was found that a lot of work has been reported on pressure sensors for sensing applications; but hitherto less work has been reported on 3D printing of smart foot sensors using bioprinting. Figure 9.3 shows the research gap between foot sensors and biomedical applications. Based on the literature it has been ascertained that less work has been reported on the monitoring of

Tab. 9.1: List of key terms based on keyword foot sensor and biomedical applications.

S. no.	Terms	Number of occurrences	Relevance scores
1	Actuator	3	0.2532
2	Ankle	3	0.7103
3	Approach	5	0.6545
4	Architecture	4	1.4724
5	Change	5	0.9782
6	Data	6	0.2921
7	Decrease	4	1.7458
8	Development	5	0.7285
9	Disease	3	1.0328
10	Effect	6	0.6218
11	Experimental result	3	0.7578
12	Fabrication	6	1.5424
13	Fbg	3	1.7449
14	Foot pressure	3	0.3252
15	Human	5	0.9993
16	Increase	4	1.5445
17	KPa	4	1.6501
18	Motion	3	0.686
19	Novel	3	0.8354
20	Patient	4	0.4623
21	Robot	8	0.7593
22	Sensitivity	5	2.2839
23	Sensor array	4	2.8164
24	Simulation	4	0.8022
25	Support	4	0.4321
26	Time	6	0.4244
27	Use	5	0.3948
28	Validation	3	1.0492

foot pressure sensors in the ISM band by using Bluetooth/Wi-Fi range as an IoT-based solution. This study highlights the sensing capability in terms of ε_r of 3D-printed smart foot sensors for continuous health monitoring of the human feet.

9.3 Materials and methodology

The methodology adopted for developing a smart foot sensor using 3D bioprinting is shown in Fig. 9.4. The first step was to design the CAD model of the foot sensor in solid works as per the calculations, using a certain set of equations (9.1)–(9.8) (Singh et al. 2022). Further, a commercially available silicone gel (make-astral adhesive) was used to 3D print the designed sensor using the Allevi-2 bioprinter, as shown in Fig. 9.5. The silicone gel was filled into a 10 mL plastic syringe, and a 21-gauge plastic nozzle having an

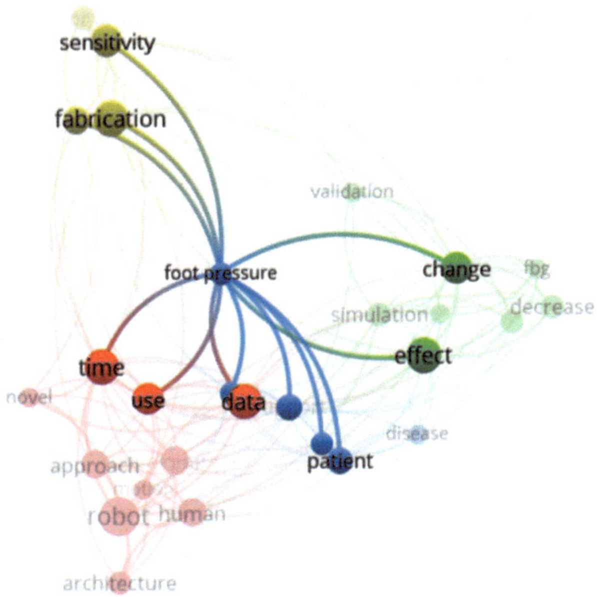

Fig. 9.3: Research gaps (based on Fig. 9.2).

outer diameter of 0.81 mm and inner diameter of 0.51 mm was used for smooth flow of material. Next, the syringe was inserted into the first extruder. The silicone gel was extruded at certain printer settings through literature review, such as infill type-grid pattern, infill distance of 1 mm, and print speed of 6 mm/s. A ring resonator test was performed on a 3D-printed part for the calculation of insertion loss (S_{21}) parameters.

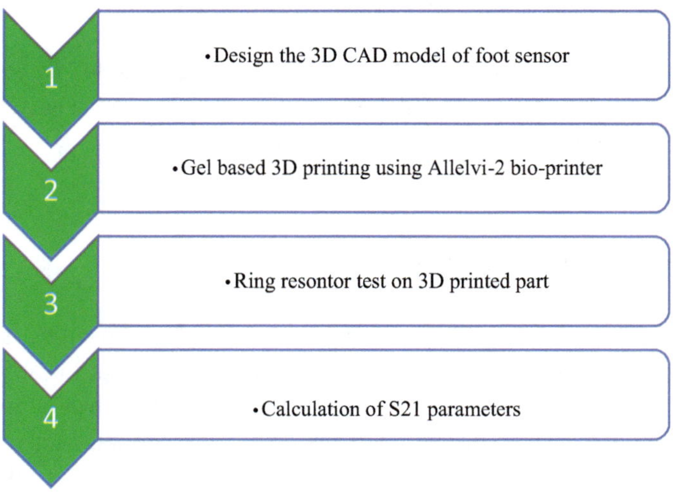

Fig. 9.4: Adopted methodology.

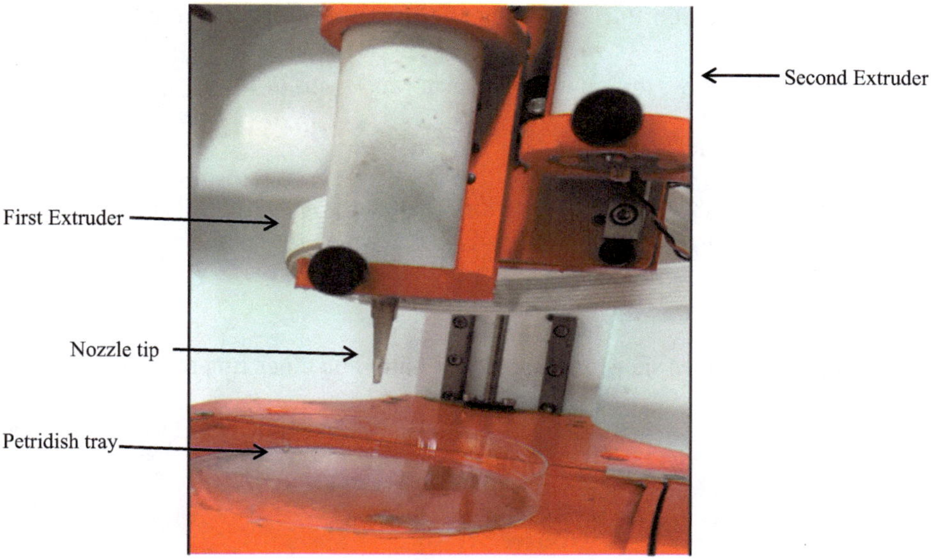

Fig. 9.5: Allelvi-2 bioprinter.

9.4 Results and discussion

9.4.1 Ring resonator test

To perform the ring resonator test, $\varepsilon_r = 11.7$ was initially assumed for silicone from the literature, and HFSS software was used to perform a ring resonator test. Fig. 9.6 shows the various RF peaks. Equations (9.1)–(9.8) (Singh et al. 2022) were used to design the ring resonator. The calculated values obtained through this test are shown in Tab. 9.2. The calculated S_{21} parameters and the design of the substrate using HFSS software are shown in Fig. 9.6.

Equation (9.1) shows the value of the effective dielectric constant (ε_{eff}):

$$\varepsilon_{\text{eff}} = \left[\frac{\epsilon_r + 1}{2}\right] + \left[\left(\frac{\epsilon_r - 1}{2}\right)\left(1 + 12\frac{h}{w}\right)^{-0.5}\right] \tag{9.1}$$

The w and h in equation (9.1) represent the feed line width and substrate height, respectively.

Equation (9.2) shows the value of the guided wavelength (λ_g)

$$\lambda_g = \frac{c}{\sqrt{(\varepsilon_{\text{eff}})}} * \frac{1}{f} \tag{9.2}$$

The c and f in equation (9.2) represent the light speed and resonant frequency, respectively.

Equation (9.3) shows the value of the mean ring radius (R):

$$2\pi R = n\lambda_g \tag{9.3}$$

The n in equation (9.3) represents the resonance order.

Equation (9.4) shows the value of the feed line length (l):

$$l = \lambda_{g/4} \tag{9.4}$$

Equations (9.5) and (9.6) show the value of the outer and inner ring radii (R_o and R_i).

$$R_o = R + \frac{w}{2} \tag{9.5}$$

$$R_i = R - \frac{w}{2} \tag{9.6}$$

Equation (9.7) shows the value of the coupling gap (C_g):

$$C_g = 0.411h \left[\frac{\epsilon_{eff} + 0.30}{\epsilon_{eff} - 0.259}\right] \times \left[\frac{w/h + 0.262}{w/h + 0.818}\right] \tag{9.7}$$

Equation (9.8) shows the value of the substrate length (L):

$$L = 2\times\ l + 2\times\ C_g + 2\ \times R_0 \tag{9.8}$$

Tab. 9.2: Parameters calculated during a simulation on HFSS.

Parameters	**For $\varepsilon_r = 11.7$**
	Values (mm)
Effective dielectric constant (ε_{eff})	7.758
Mean ring radius (R)	15.820
Length of feed line (l)	31.130
Feed line width (w)	2.234
The outer ring radius (R_o)	20.934
The inner ring radius (R_i)	18.700
Coupling gap (C_g)	0.74543
Substrate length (L)	100.000
Substrate width (W)	101.863

The observed S_{21} parameters for $\varepsilon_r = 11.7$ were −33.70 dB, observed at RF = 2.1429 GHz, and −40.3170 dB, observed at RF = 2.6327 GHz, as shown in Fig. 9.6.

Output plots and design of ring resonator	ε_r=11.7
S21	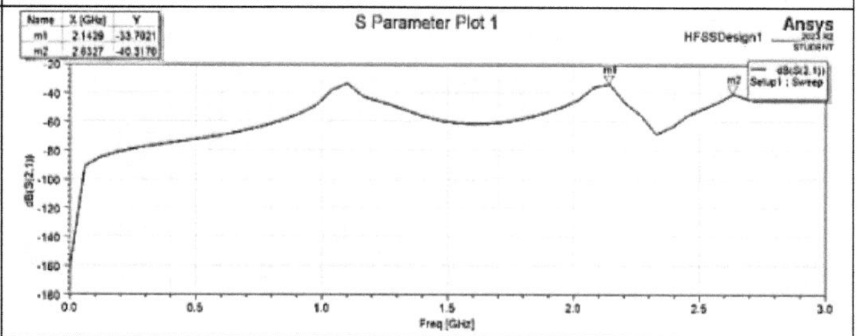
Design	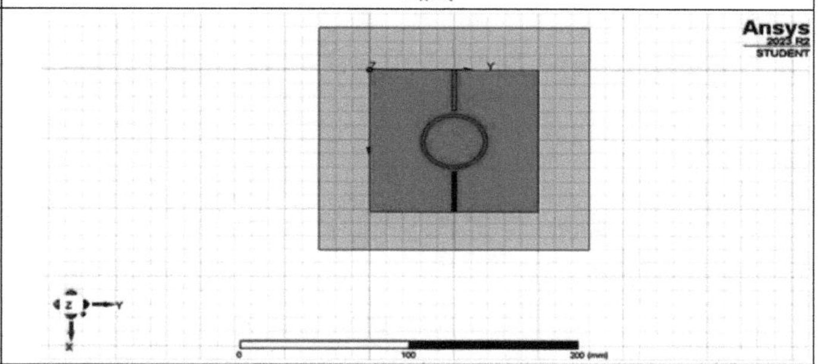

Fig. 9.6: S_{21} parameters and substrate design for $\varepsilon_r = 11.7$.

9.4.2 RF characteristics of the 3D-printed foot sensor

The S_{21} parameters (Fig. 9.7) for the designed ring resonator were obtained through a vector network analyzer (VNA) and it was calculated for a frequency range of 1–3 GHz. It was observed that the ring resonator resonates at 2.38136 GHz and the measured S_{21} parameter corresponding to the resonant frequency was −20.938 dB. Further, ε_r and tan δ have been calculated using equations of ring resonators as per past literature. The observed values of ε_r and tan δ were 0.5987 and 0.000418.

It has been observed that the fabricated sensor (3D bioprinted smart insole foot sensor) is capable of identifying the shoe articles during packaging in terms of variation of ε_r that can be configured with a nearby Bluetooth device, reducing the identification time and is a ready-to-use solution for the shoe industry.

9.5 Conclusions

The simulated results of S_{21} parameters for $\varepsilon_r = 11.7$ were −33.7021 dB at RF = 2.1429 GHz and −40.3170 dB at RF = 2.6327 GHz. It has been observed that the designed sensor can transmit the signal within a range of the ISM band for the continuous monitoring of

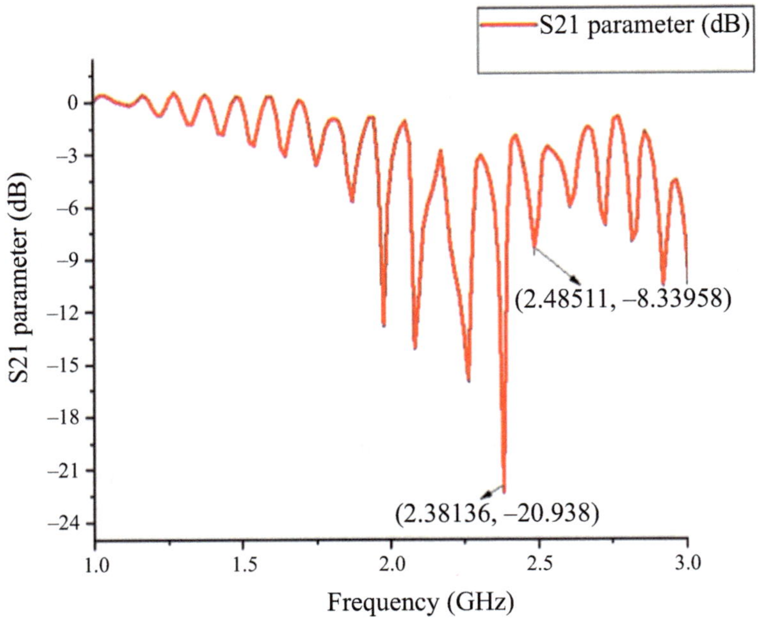

Fig. 9.7: Variation in S_{21} versus frequency for output of ring resonator.

pressure, even in remote areas. The RF characteristics of the foot sensor show that the sensor can be proposed for sports practice as well as for biomedical applications. The foot sensor is capable of identifying the shoe articles during packaging, in terms of variation of ε_r that may be configured with a nearby Bluetooth device, reducing the identification time and is a ready-to-use solution for the shoe industry.

9.6 Future scope

This work uses a silicone gel for the fabrication of a 3D bioprinted smart insole foot sensor. Future work may be focused on the use of piezoelectric material, which changes its phase when the load (foot pressure) is applied to it, resulting in a variation of ε_r that will help to diagnose the continuous pressure distribution. Also, there is scope for smart foot sensors in spy applications, as a shift in the ε_r parameter will shift the frequency of the incoming signal, indicating the requirement for necessary aid during an emergency. Smart foot sensors have scope to reduce the bad odor from feet, as a smart 4D drug material may be inserted in the insole, which can release the drug due to the application of foot pressure to avoid any skin allergies.

References

Aqueveque P, Osorio R, Pastene F, Saavedra F, Pino E. Capacitive sensors array for plantar pressure measurement insole fabricated with flexible PCB. In 2018 40th Annual International Conference of the IEEE Engineering in Medicine and Biology Society (EMBC). 2018 Jul;18:4393–96. IEEE.

Avedesian JM, Covassin T, Dufek JS. The influence of sport-related concussion on lower extremity injury risk: A review of current return-to-play practices and clinical implications. International Journal of Exercise Science. 2020;13(3):873.

Barkley RM, Bumgarner MR, Poss EM, Senchina DS. Physiological versus perceived foot temperature, and perceived comfort, during treadmill running in shoes and socks of various constructions. American Journal of Undergraduate Research. 2011;10(3):7–14.

Chatwin KE, Abbott CA, Boulton AJ, Bowling FL, Reeves ND. The role of foot pressure measurement in the prediction and prevention of diabetic foot ulceration – A comprehensive review. Diabetes/ Metabolism Research and Reviews. 2020 May;36(4):e3258.

Chen J, Dai Y, Kang S, Xu L, Gao S. A concurrent plantar stress sensing and energy harvesting technique by piezoelectric insole device and rectifying circuitry. IEEE Sensors Journal. 2021 Mar 8;21(23):26364–72.

Cicirelli G, Impedovo D, Dentamaro V, Marani R, Pirlo G, D'Orazio TR. Human gait analysis in neurodegenerative diseases: A review. IEEE Journal of Biomedical and Health Informatics. 2021 Jun 28;26(1):229–42.

Federation ID. IDF Diabetes Atlas, (9th Edition). 2019.

Husain M, Singh R, Pabla BS. A review on 3d printing of partially absorbable implants. Journal of the Institution of Engineers (India): Series C. 2023a Jul;23:1–20.

Husain M, Singh R, Pabla BS. On 3D printing of pvdf composite-based sensors for biomedical applications. National Academy Science Letters. 2023b Jun;22:1–6.

Husain M, Singh R, Pabla BS. On compressive properties of 3D printed polyvinylidene fluoride composite for implant applications. Journal of Thermoplastic Composite Materials. 2023c Sep;5:08927057231200005.

Husain M, Singh R, Pabla BS. On PVDF composite as partially absorbable smart implants. Proceedings of the Institution of Mechanical Engineers, Part H: Journal of Engineering in Medicine. 2023d Apr;237 (4):517–26.

Jor A, Das S, Bappy AS, Rahman A. Foot plantar pressure measurement using low-cost force sensitive resistor (FSR): Feasibility study. Journal of Scientific Research. 2019 Sep 1;11(3):311–19.

Jung Y, Lee W, Jung K, Park B, Park J, Ko J, Cho H. A highly sensitive and flexible capacitive pressure sensor based on a porous three-dimensional PDMS/microsphere composite. Polymers. 2020 Jun 24;12 (6):1412.

Leal-Junior AG, Marques C. Diaphragm-embedded optical fiber sensors: A review and tutorial. IEEE Sensors Journal. 2020 Dec 3;21(11):12719–33.

Lekkala J. Plantar shear stress measurements – A review. Clinical Biomechanics. 2014 May 1;29(5):475–83.

Minty E, Bray E, Bachus CB, Everett B, Smith KM, Matijevich E, Hajizadeh M, Armstrong DG, Liden B. Preventative sensor-based remote monitoring of the diabetic foot in clinical practice. Sensors. 2023 Jul 27;23(15):6712.

Moccia S, Nucci L, Spagnuolo C, d'Apuzzo F, Piancino MG, Minervini G. Polyphenols as potential agents in the management of temporomandibular disorders. Applied Sciences. 2020 Jul 31;10(15):5305.

Oliveira SS, Pannuti CM, Paranhos KS, Tanganeli JP, Laganá DC, Sesma N, Duarte M, Frigerio ML, Cho SC. Effect of occlusal splint and therapeutic exercises on postural balance of patients with signs and symptoms of temporomandibular disorder. Clinical and Experimental Dental Research. 2019 Apr;5 (2):109–15.

Prompers L, Huijberts M, Schaper N, Apelqvist J, Bakker K, Edmonds M, Holstein P, Jude E, Jirkovska A, Mauricio D, Piaggesi A. Resource utilization and costs associated with the treatment of diabetic foot ulcers. Prospective data from the Eurodiale Study. Diabetologia. 2008 Oct;51:1826–34.

Rajendran D, Ramalingame R, Palaniyappan S, Wagner G, Kanoun O. Flexible ultra-thin nanocomposite based piezoresistive pressure sensors for foot pressure distribution measurement. Sensors. 2021 Sep 10;21(18):6082.

Ramirez-Bautista JA, Huerta-Ruelas JA, Chaparro-Cárdenas SL, Hernández-Zavala A. A review in detection and monitoring gait disorders using in-shoe plantar measurement systems. IEEE Reviews in Biomedical Engineering. 2017 Aug 30;10:299–309.

Razak AH, Zayegh A, Begg RK, Wahab Y. Foot plantar pressure measurement system: A review. Sensors. 2012 Jul 23;12(7):9884–912.

Samarentsis AG, Makris G, Spinthaki S, Christodoulakis G, Tsiknakis M, Pantazis AK. A 3D-printed capacitive smart insole for plantar pressure monitoring. Sensors. 2022 Dec 12;22(24):9725.

Shi J. Textile-based ferroelectric for the foot pressure sensor. In2019 IEEE International Conference on Flexible and Printable Sensors and Systems (FLEPS). 2019 Jul;8:1–3. IEEE.

Singh R, Kumar S, Singh AP, Wei Y. On comparison of recycled LDPE and LDPE–bakelite composite based 3D printed patch antenna. Proceedings of the Institution of Mechanical Engineers, Part L: Journal of Materials: Design and Applications. 2022 Apr;236(4):842–56.

Tao J, Dong M, Li L, Wang C, Li J, Liu Y, Bao R, Pan C. Real-time pressure mapping smart insole system based on a controllable vertical pore dielectric layer. Microsystems & Nanoengineering. 2020 Aug 10;6(1):62.

Yang Y, Gao W, Zhao Z. Research on gait cycle recognition with plantar pressure sensors. Proceedings of the 4th International Conference on Computer Science and Application Engineering. 2020 Oct;20:1–5.

Ranvijay Kumar, Rupinder Singh, Vishal Thakur*, Deepika Kathuria, Seema Ramniwas

10 3D-printed stockings for controlled drug delivery

Abstract: *Salix alba* as a drug has many medicinal benefits like pain relief, anti-inflammatory effects, fever reduction, migraine relief, relief in gastrointestinal discomfort, and menstrual pain relief. This research study aims to develop the *Salix alba* powder-reinforced polycaprolactone (PCL)–polyvinyl alcohol (PVA) composite solution used for the controlled release for additive manufacturing of stockings. *Salix alba* was reinforced with various compositions of PCL and PVA to manufacture stockings for drug release at a controlled rate. PCL–PVA blended with chloroform to form a gel solution and *Salix alba* was incorporated by 5 wt.% in each composition. The drug release for each fabricated composite with varied compositions was tested by using UV spectroscopy. From the UV spectroscopy, it was observed that the controlled drug release depended on various parameters such as drug variation of the PVA loading in the PCL matrix and uniformity of the *Salix alba* particle distribution.

Acknowledgment: The authors are thankful to the Science and Engineering Research Board for funding under TAR/2021/000126.

***Corresponding author: Vishal Thakur**, Department of Mechanical Engineering, Chandigarh University, Mohali, Punjab, India, e-mail: vishal2104.thakur@gmail.com
Deepika Kathuria, Seema Ramniwas, University Centre for Research and Development, Chandigarh University, Mohali, Punjab, India
Ranvijay Kumar, University Centre for Research and Development, Chandigarh University, Mohali, Punjab, India; Department of Mechanical Engineering, National Institute of Technical Teachers Training and Research, Chandigarh 160019, India
Rupinder Singh, Department of Mechanical Engineering, National Institute of Technical Teachers Training and Research, Chandigarh 160019, India

https://doi.org/10.1515/9783111081632-010

Graphical abstract:

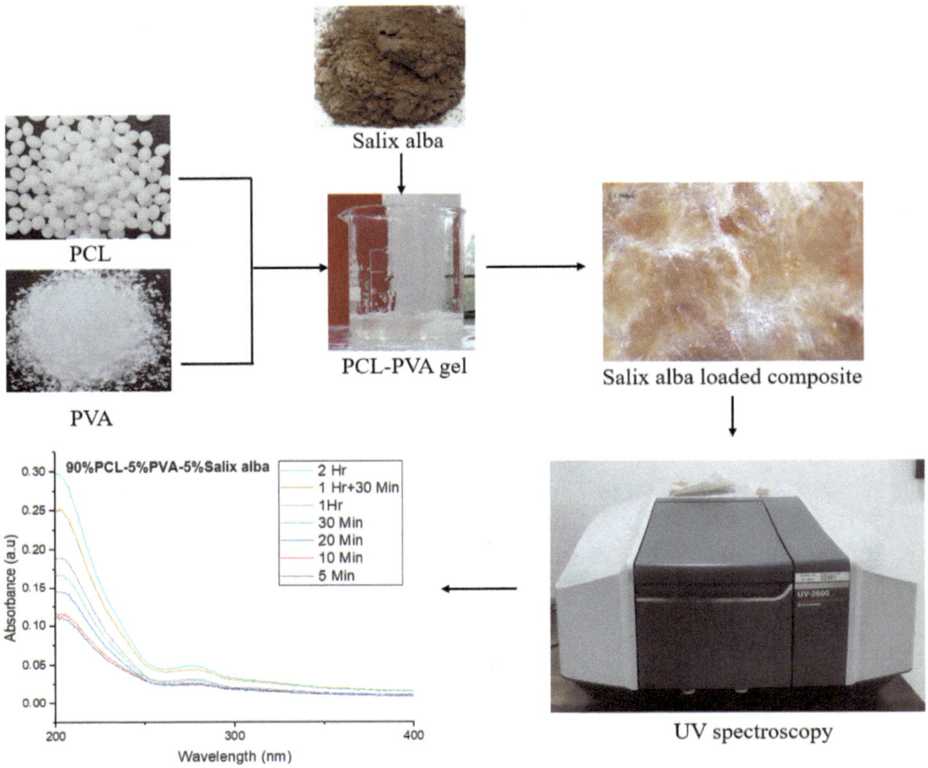

Keywords: PVA, PLA, *Salix alba*, 3D-printed stockings, drug delivery

10.1 Introduction

3D printing has evolved with the manufacturing of day-to-day utilities in almost every application domain. In the healthcare industry, 3D printing plays an important role in the manufacturing of biomedical implants, scaffolding, podiatric medicine, and pharmacy goods (Placone and Engler 2018, Ahangar et al. 2019, Beg et al. 2020). In the recent development, stockings are being used for the relaxation and relief of stressed tissue and bones. 3D printing can manufacture stockings for certain biomedical applications. However, the materials of the 3D-printed stocking are the major area of concern which must be developed in such as way that should act as the multifunctional components. The functionality of the 3D-printing stocking is dependent upon what reinforcement is being used with the polymer matrix. *Salix alba* bark is one of the plant extracts that contains salicin (similar structure to aspirin) which has excellent medicinal uses, especially in rain relieving or wound healing applications (Sulaiman et al. 2013, Kenstavičienė et al. 2009, Qureshi et al. 2015). The medicinal uses of *Salix alba* include pain relief, anti-inflammatory effects, fever reduction, migraine relief, gastrointestinal discomfort, menstrual pain relief, and topical applications in ointments or creams to alleviate localized pain, such as muscle and joint pain. The success of these types of medicinal materials depends on the control of the drug release during the treatment phase. Drugs should be released at a controlled rate to cure the issues. PVA is one of the synthetic polymers that can be dissolved in water at ambient temperature. It has been reported by previous investigators that the PVA is being used as the host materials to be utilized in controlled drug delivery in wound healing applications (Sung et al. 2010, Saraiva et al. 2023, Venkataprasanna et al. 2020, You et al. 2004). PVA is one of the most useful polymers in pharmacy applications where capsules are tablets are being 3D-printed for controlled drug loading and release. Matijašić et al. (2019) have reported the design and manufacturing of multicompartmental capsules of PVA materials prepared by the 3D-printing process. Similarly, previous studies have reported 3D printing of the PVA-based materials in capsular devices, personalized drug delivery, oral solid dosages of drugs, and so on (Cotabarren and Gallo 2020, Jain et al. 2018, Gioumouxouzis et al. 2017, Goyanes et al. 2016).

In many cases, the scaffolds are loaded with drugs and medicines to promote the fast curing of injuries as well as to reduce the chances of infections due to microbial growth. PCL is one of the most reported thermoplastics in the 3D printing of biomedical scaffolds in tissue engineering applications. The PCL thermoplastic has the advantage of good thermoresponsive behavior due to that it acts as a shape memory polymer. The good shape memory behavior of the PCL makes it an eligible candidate for the manufacturing of smart scaffolds responsive to temperature. The previous studies have revealed the utilities of the PCL-based matrix for tissue engineering, bone tissue engineering bladder tissue engineering, and so on (Patricio et al. 2013, Oh et al. 2007, Shor et al. 2009; Baker et al. 2009). Some of the previous studies have reported the use of the PCL as the matrix materials in 3D printing for tissue engineering, vascularized bone grafts, cartilage mimicking, and so on (Jiao et al. 2019, Temple et al. 2014, Ma et al. 2019, Olubamiji et al. 2016).

The literature review has suggested that PCL is commonly used in the manufacturing of biomedical scaffolds. PVA is being used for the controlled delivery of drugs for remedial purposes. Some of the research studies have highlighted the use of *Salix alba* materials in wound healing applications. But hitherto, very less studies have been reported for the 3D printing of the stocking having controlled drug release capabilities. This chapter is focused on the preparation of the *Salix alba*-loaded PCL–PVA polymer matrix for possible 3D printing of the stocking having controlled drug release capabilities. This study includes the investigations of the drug release from the PCL–PVA–*Salix alba*-based polymer composites based upon the UV–visible spectroscopy results.

10.2 Background

The background of the study has been checked by using the available data on www.webofknowledge.com website. In the previous 17 years (2007–2023), there are 882 results observed with keywords of "controlled drug delivery" and "3D printing." Figure 10.1 shows the maximum number of research papers reported in 2021 and the research work increases at a significant rate every year related to controlled drug delivery using 3D-printing techniques.

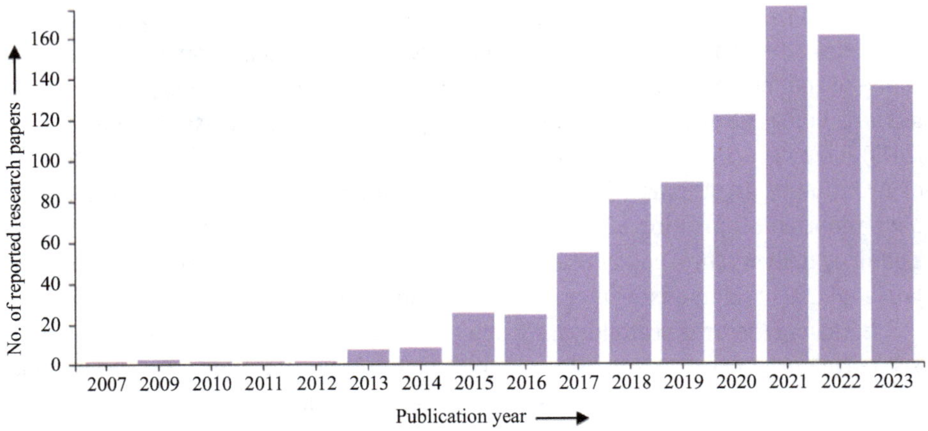

Fig. 10.1: Year-wise count of published research papers for controlled drug delivery using 3D-printing techniques (Fig. Courtesy: www.webofknowledge.com).

The evaluation of gaps in published literature was done by analyzing the database obtained from Web of Science with VosViewer software. Out of 20,812 terms among 882 published literature, 181 terms were filtered by keeping 20 occurrences of terms. Figure 10.2 shows the relationship diagram which is prepared by using the 24 most relevant terms and also represents the connection between the 3D-printing process and fabrication of medical products used for drug delivery applications. Figure 10.3 shows that more investigations may be performed for the fabrication of 3D-printed medical devices or capsules for drug release at a controlled rate. The selected terms, their occurrences, and their relevance score are shown in Tab. 10.1.

10.3 Materials and methods

The complete procedure of experiments is detailed in Fig. 10.4. In the first stage, the bark of *Salix alba* from a 5-year-old tree was taken and the moisture was removed by keeping it in a hot air oven. In the next stage, the bark of *Salix alba* was processed on

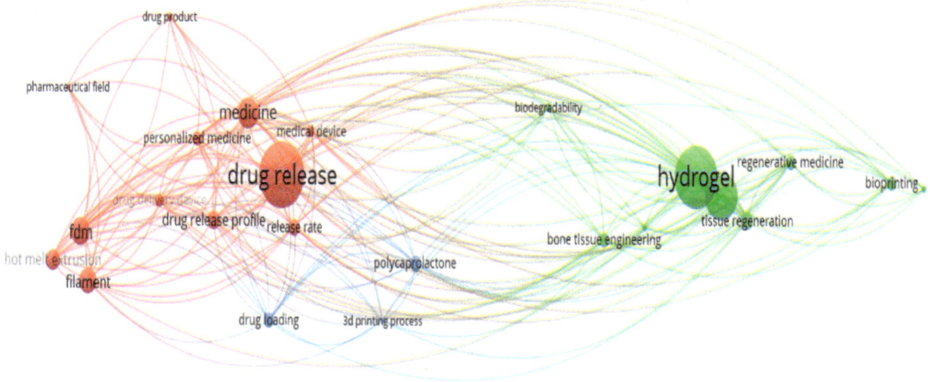

Fig. 10.2: Relationship diagram for controlled drug delivery using a 3D-printing process.

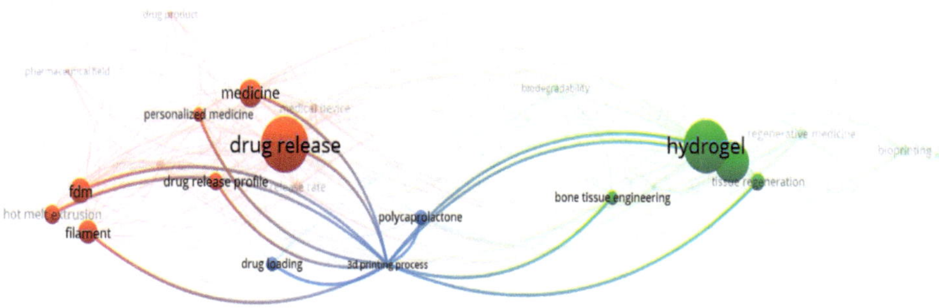

Fig. 10.3: Gaps in the reported publications related to 3D printing and drug delivery.

Tab. 10.1: Selected terms, their occurrences, and relevance scores.

ID	Terms	Occurrences	Relevance scores
1	3D bioprinting	18	3.3496
2	3D-printing process	16	0.0709
3	Biodegradability	23	0.6378
4	Bioprinting	34	2.932
5	Bone tissue engineering	34	0.7815
6	Cell proliferation	18	0.9225
7	Drug delivery device	30	0.6716
8	Drug loading	38	0.4411
9	Drug product	20	0.7077
10	Drug release	158	0.2647
11	Drug release profile	45	0.4075
12	FDM	66	1.1488
13	Filament	62	1.2093
14	Hot melt extrusion	49	1.2898

Tab. 10.1 (continued)

ID	Terms	Occurrences	Relevance scores
15	Hydrogel	150	1.4247
16	Medical device	29	0.1383
17	Medicine	74	0.4214
18	Personalized medicine	36	0.3962
19	Pharmaceutical field	16	1.1407
20	Polycaprolactone	40	0.1088
21	Regenerative medicine	35	1.9062
22	Release rate	41	0.2556
23	Tissue engineering	114	1.5035
24	Tissue regeneration	42	1.8698

the ball milling machine to make it in the form of powder (size: 40–50-μm size). In the next stage, the 12 wt.% PCL was dissolved in the acetone for gel formation. The 12-wt PVA was dissolved in water to form the gel. In the next stage, gels of both PCL and PVA are mixed on a magnetic stirrer to form the composite gels. In the next stage, the *Salix alba* powder was mixed with PLA–PVA-blended gel and then it was waited to remove the moisture from the PLA–PVA–*Salix alba* composite gel. In the further stage, the drug release of the *Salix alba* was measured using the UV–visible spectroscopy setup.

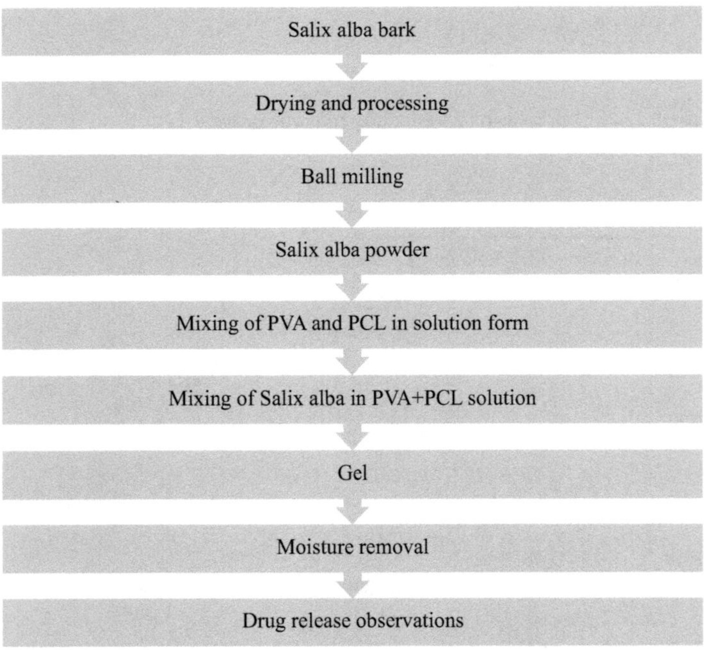

Fig. 10.4: Methodology for preparation of PCL–PVA–*Salix alba*-based composites.

10.4 Experimentation

In the experimentation of composite formation, there are four different percentage loading of PVA (5%, 10%, 15%, and 20%) induced in PCL. A constant amount of 5% *Salix alba* powder was loaded in each sample as shown in Tab. 10.2. To measure the drug release capabilities of the *Salix alba* from the PCL–PVA–*Salix alba* composite blend, the composite block was put in the fresh distilled water in the laboratory beaker. After this, the sample was taken from the weaker, and absorbance was noted on a UV–visible spectrophotometer. The same process was repeated after a certain interval, till it got the constant absorbance on the UV–visible spectrophotometer. The concentration of the released drug was evaluated with the help of a standard curve of drugs prepared from different unknown concentrations. Absorption spectra were observed in the range of 200–400 nm wavelength. The drug release of the *Salix alba* was noted at intervals of 5, 10, 20, 30, 60, 90, and 120 min.

Tab. 10.2: Composition of *Salix alba*-loaded PCL–PVA composites.

Samples	PCL (wt.%)	PVA (wt.%)	Salix alba (wt.%)
1	90	5	5
2	85	10	5
3	80	15	5
4	75	20	5

10.5 Results and discussion

Figure 10.5 shows the absorption spectra of *Salix alba* powder. It has been observed that in the range of 200–250 nm wavelength, the release of the *Salix* was maximum, and almost negligible release was observed in the range of 300–400 nm wavelength.

The release of the *Salix alba* from the PCL–PVA polymer matrix was observed in the specified interval of time. The trend of the *Salix alba* release confirmed that the drug release from the composite matrix was steady till 2 h of the observation. There is no nonuniform absorption spectra were noted which confirms that the mixing of either PLA with PVA or the mixing of *Salix alba* with PCL–PVA gel is uniform. The UV–visible spectra of 90% PCL–5% PVA–5% *Salix alba* sample are given in Fig. 10.6.

In the case of a composite sample composed of 85% PCL–10% PVA–*Salix alba* (Fig. 10.7), the drug release was observed higher than the 90% PCL–5% PVA–*Salix alba* sample in 2 h with a significant margin. Increasing the PVA loading in the PCL–*Salix alba* mixture has increased the drug release capabilities. This is because the PVA is a water-soluble polymer and increasing its proportion has increased the potential drug release.

Figure 10.8 shows the UV–visible spectra of the drug release of the *Salix alba* from 80% PCL–15% PVA–5% *Salix alba*-based composite. As per the uniformity of the

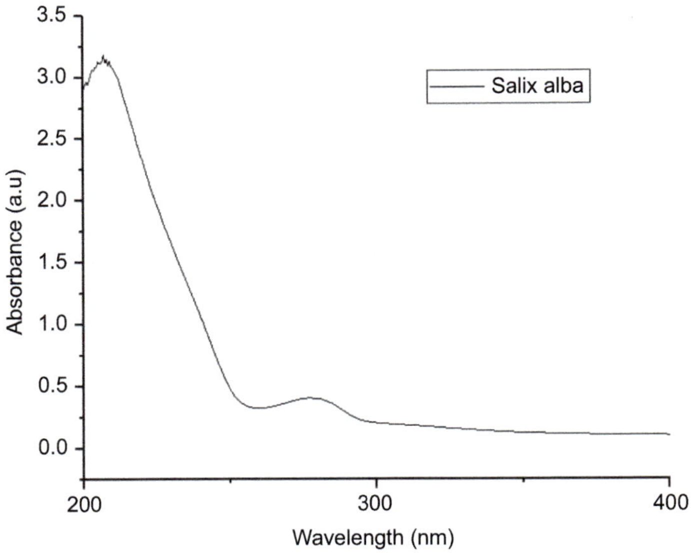

Fig. 10.5: Absorption spectrum of *Salix alba*.

mixing/blending of powder concerns, there are some nonuniform behaviors of the release pattern were observed. It has been observed that the abruption of 80% PCL–15% PVA–5% *Salix alba*-based composite sample is higher as compared to the 90% PCL–5% PVA–5% *Salix alba* and 85% PCL–10% PVA–5% *Salix alba* samples.

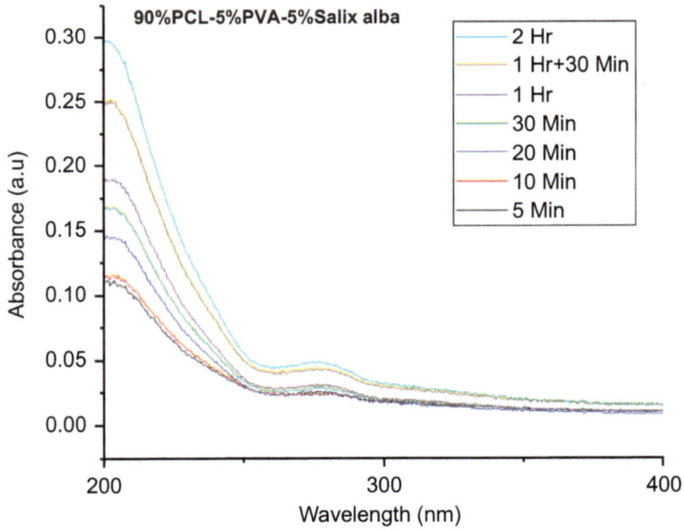

Fig. 10.6: Absorption spectra of 90% PCL–5% PVA–*Salix alba*-based composite.

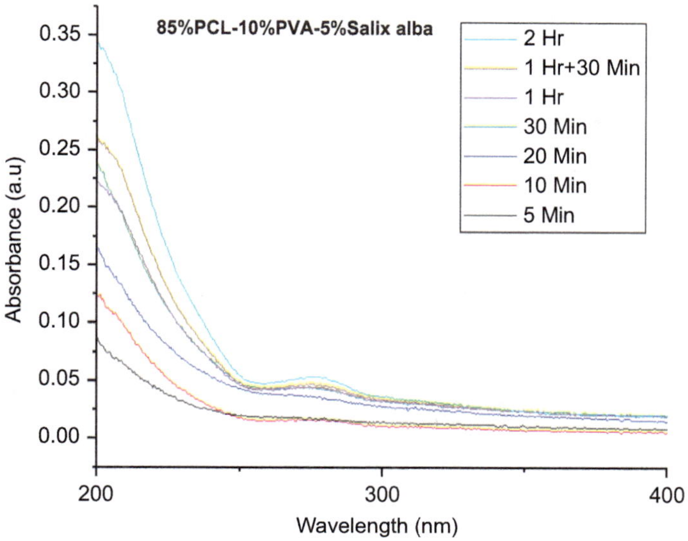

Fig. 10.7: Absorption spectra of 85% PCL–10% PVA–5% *Salix alba*-based composite.

In the case of a composite sample made up of Fig. 10.9 absorption spectra of 85% PCL–10% PVA–5% *Salix alba*, the uniform release of the drug release was observed. This may be because uniform mixing of the *Salix alba* in the case of sample 4 was noted. Also, the release of the 85% PCL–10% PVA–5% *Salix alba*-based composite was observed higher as compared to other samples. Overall, the composites prepared by the present processing method have good control over the drugs. The proposed composite may be used for applications such as additive manufacturing of the stockings.

The micrographs were captured by using an optical microscope for various compositions of PCL (75, 80, 85, and 90 wt.%), PVA (5, 10, 15, and 20 wt.%), and *Salix alba* (5 wt.%)-based composites as shown in Fig. 10.10, and PCL and PVA were used as drug carriers. From the micrographs, it was observed that with the increase in PVA composition, the surface texture became rough and porous. It has been noticed that the drug release capabilities are directly proportional to the loading of the PVA in the PCL matrix. Previous studies have suggested that the increase in porosity leads to increased drug release capabilities (Braunecker et al. 2004). So, the drug composition should be optimized for the release of a drug at a controlled release rate.

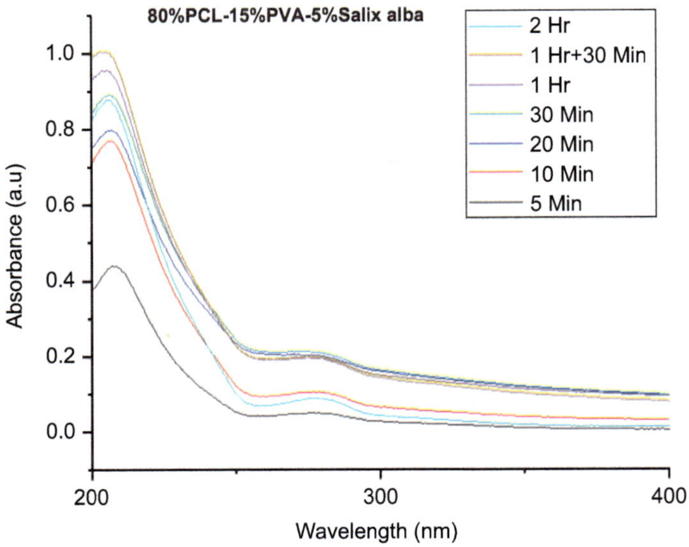

Fig. 10.8: Absorption spectra of 80% PCL–15% PVA–5% *Salix alba*-based composite.

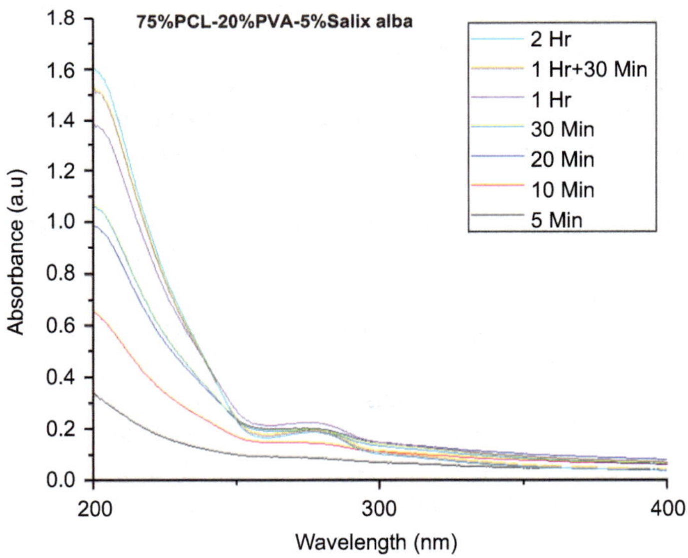

Fig. 10.9: Absorption spectra of 75% PCL–20% PVA–5% *Salix alba*-based composite.

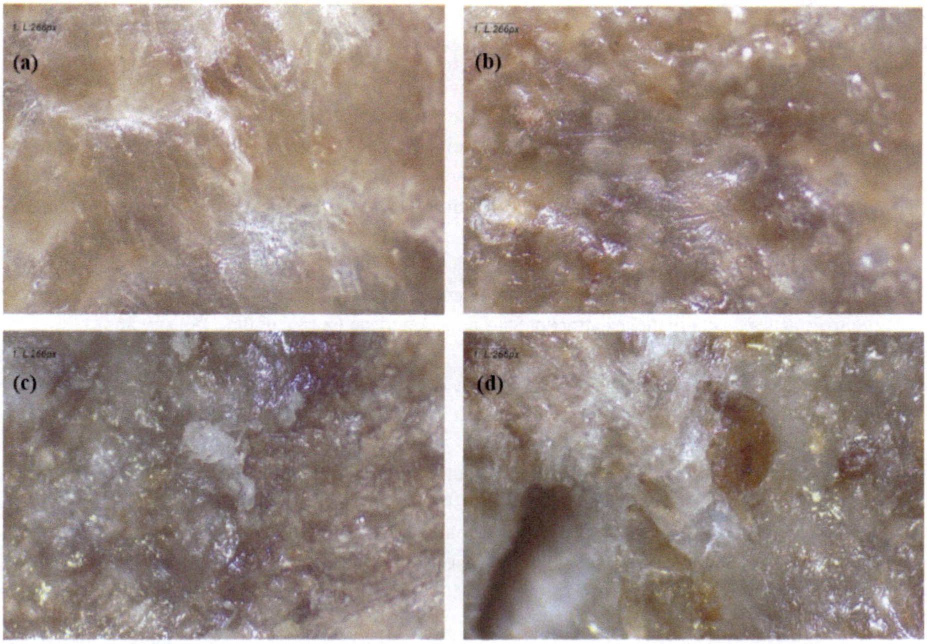

Fig. 10.10: Micrograph for (a) 90% PCL–5% PVA–5% *Salix alba*; (b) 85% PCL–10% PVA–5% *Salix alba*; (c) 80% PCL–15% PVA–5% *Salix alba*; and (d) 75% PCL–20% PVA–5% *Salix alba*-based composite.

10.6 Conclusions

The *Salix alba* drug was reinforced with PCL–PVA gel solution. The observations have shown that PCL–PVA–*Salix alba*-based composite solutions have been successfully developed. As compared to the release of *Salix alba* powder in distilled water, the drug release of PCL–PVA–*Salix alba*-based composites is comparatively slow. It has been observed that the drug release of 90% PCL–5% PVA–5% *Salix alba*-based composites solution was with minimum rate whereas the drug release of 75% PCL–20% PVA–5% *Salix alba*-based composite solution was with maximum rate. Overall, the drug release was increased with the increasing proportion of PVA in the composite. The PVA material is water soluble material, so when it comes in contact with an aqueous medium, it exhibits an enhanced drug release potential within the fabricated composite. If the drug solubility is low, it takes hours in the process of drug release. The results suggest that higher and lower concentration of PVA with PCL was used for the fabrication of high-soluble drugs and low-soluble drugs, respectively.

References

Placone JK, Engler AJ. Recent advances in extrusion-based 3D printing for biomedical applications. Advanced Healthcare Materials. 2018 Apr;7(8):1701161.

Ahangar P, Cooke ME, Weber MH, Rosenzweig DH. Current biomedical applications of 3D printing and additive manufacturing. Applied Sciences. 2019 Apr 25;9(8):1713.

Beg S, Almalki WH, Malik A, Farhan M, Aatif M, Rahman Z, Alruwaili NK, Alrobaian M, Tarique M, Rahman M. 3D printing for drug delivery and biomedical applications. Drug Discovery Today. 2020 Sep 1;25 (9):1668–81.

Sulaiman GM, Hussien NN, Marzoog TR, Awad HA. Phenolic content, antioxidant, antimicrobial, and cytotoxic activities of ethanolic extract of Salix alba. American Journal of Biochemistry and Biotechnology. 2013 Jan 1;9(1):41–46.

Kenstavičienė P, Nenortienė P, Kiliuvienė G, Ževžikovas A, Lukošius A, Kazlauskienė D. Application of high-performance liquid chromatography for research of salicin in bark of different varieties of Salix. Medicina. 2009 Aug;45(8):644.

Qureshi MA, Khatoon F, Ahmed S. An overview on wounds, their issues and natural remedies for wound healing. Biochemistry and Physiology. 2015;4(3):2–9.

Sung JH, Hwang MR, Kim JO, Lee JH, Kim YI, Kim JH, Chang SW, Jin SG, Kim JA, Lyoo WS, Han SS. Gel characterisation and in vivo evaluation of minocycline-loaded wound dressing with enhanced wound healing using polyvinyl alcohol and chitosan. International Journal of Pharmaceutics. 2010 Jun 15;392 (1–2):232–40.

Saraiva MM, Campelo MD, Camara Neto JF, Lima AB, Silva GD, Dias AT, Ricardo NM, Kaplan DL, Ribeiro ME. Alginate/polyvinyl alcohol films for wound healing: Advantages and challenges. Journal of Biomedical Materials Research Part B: Applied Biomaterials. 2023 Jan;111(1):220–33.

Venkataprasanna KS, Prakash J, Vignesh S, Bharath G, Venkatesan M, Banat F, Sahabudeen S, Ramachandran S, Venkatasubbu GD. Fabrication of chitosan/PVA/GO/CuO patch for potential wound healing application. International Journal of Biological Macromolecules. 2020 Jan 15;143:744–62.

You Y, Park WH, Ko BM, Min BM. Effects of PVA sponge containing chitooligosaccharide in the early stage of wound healing. Journal of Materials Science: Materials in Medicine. 2004 Mar;15:297–301.

Matijašić G, Gretić M, Vinčić J, Poropat A, Cuculić L, Rahelić T. Design and 3D printing of multi-compartmental PVA capsules for drug delivery. Journal of Drug Delivery Science and Technology. 2019 Aug 1;52:677–86.

Cotabarren I, Gallo L. 3D printing of PVA capsular devices for modified drug delivery: Design and in vitro dissolution studies. Drug Development and Industrial Pharmacy. 2020 Sep 1;46(9):1416–26.

Jain V, Haider N, Jain K. 3D printing in personalized drug delivery. Current Pharmaceutical Design. 2018 Nov 1;24(42):5062–71.

Mohapatra S, Kar RK, Biswal PK, Bindhani S. Approaches of 3D printing in current drug delivery. Sensors International. 2022 Jan 1;3:100146.

Gioumouxouzis CI, Katsamenis OL, Bouropoulos N, Fatouros DG. 3D-printed oral solid dosage forms containing hydrochlorothiazide for controlled drug delivery. Journal of Drug Delivery Science and Technology. 2017 Aug 1;40:164–71.

Goyanes A, Kobayashi M, Martínez-Pacheco R, Gaisford S, Basit AW. Fused-filament 3D printing of drug products: Microstructure analysis and drug release characteristics of PVA-based caplets. International Journal of Pharmaceutics. 2016 Nov 30;514(1):290–95.

Patricio T, Domingos M, Gloria A, Bártolo P. Characterisation of PCL and PCL/PLA scaffolds for tissue engineering. Procedia CIRP. 2013 Jan 1;5:110–14.

Oh SH, Park IK, Kim JM, Lee JH. In vitro and in vivo characteristics of PCL scaffolds with pore size gradient fabricated by a centrifugation method. Biomaterials. 2007 Mar 1;28(9):1664–71.

Shor L, Güçeri S, Chang R, Gordon J, Kang Q, Hartsock L, An Y, Sun W. Precision extruding deposition (PED) fabrication of polycaprolactone (PCL) scaffolds for bone tissue engineering. Biofabrication. 2009 Mar 20;1(1):015003.

Baker SC, Rohman G, Southgate J, Cameron NR. The relationship between the mechanical properties and cell behaviour on PLGA and PCL scaffolds for bladder tissue engineering. Biomaterials. 2009 Mar 1;30(7):1321–28.

Jiao Z, Luo B, Xiang S, Ma H, Yu Y, Yang W. 3D printing of HA/PCL composite tissue engineering scaffolds. Advanced Industrial and Engineering Polymer Research. 2019 Oct 1;2(4):196–202.

Temple JP, Hutton DL, Hung BP, Huri PY, Cook CA, Kondragunta R, Jia X, Grayson WL. Engineering anatomically shaped vascularized bone grafts with hASCs and 3D-printed PCL scaffolds. Journal of Biomedical Materials Research Part A. 2014 Dec;102(12):4317–25.

Ma J, Lin L, Zuo Y, Zou Q, Ren X, Li J, Li Y. Modification of 3D printed PCL scaffolds by PVAc and HA to enhance cytocompatibility and osteogenesis. RSC Advances. 2019;9(10):5338–46.

Olubamiji AD, Izadifar Z, Si JL, Cooper DM, Eames BF, Chen DX. Modulating mechanical behaviour of 3D-printed cartilage-mimetic PCL scaffolds: Influence of molecular weight and pore geometry. Biofabrication. 2016 Jun 23;8(2):025020.

Braunecker J, Baba M, Milroy GE, Cameron RE. The effects of molecular weight and porosity on the degradation and drug release from polyglycolide. International Journal of Pharmaceutics. 2004 Sep 10;282(1–2):19–34.

Vishal Thakur*, Ranvijay Kumar, Rupinder Singh

11 3D printing-based smart solutions to boost the circular economy

Abstract: A circular economy is a process of eliminating waste and using resources continuously. A circular economy is an advantageous system for achieving sustainability and resource efficiency that has recently attracted a lot of attention. Additive manufacturing (AM) is a key technology to reduce the consumption of resources and enhance business opportunities. This chapter focused on the recycling processes for waste plastic products, materials that can be recycled, contamination level in the recycling of products, and sustainability aspects for AM of recycled materials. The process of AM is the basis of the new digital manufacturing, increasing the operational and effectiveness of opportunities. A sustainable manufacturing solution that promotes efficiency and environmental sustainability related to the circular economy may be possible using AM techniques. AM using an approach based on the circular economy can offer sufficient advantages for polymer reuse as well as recycling. This chapter contains information on the recycling of thermoplastic waste, increases awareness of the circular economy, and encourages individuals or organizations to make responsible choices to reduce the impact on the environment from the waste.

***Corresponding author: Vishal Thakur**, Department of Mechanical Engineering, Chandigarh University, Mohali, Punjab, India, e-mail: vishal2104.thakur@gmail.com
Ranvijay Kumar, University Centre for Research and Development, Chandigarh University, Mohali, Punjab, India
Rupinder Singh, Department of Mechanical Engineering, Chandigarh University, Mohali, Punjab, India

https://doi.org/10.1515/9783111081632-011

Graphical abstract:

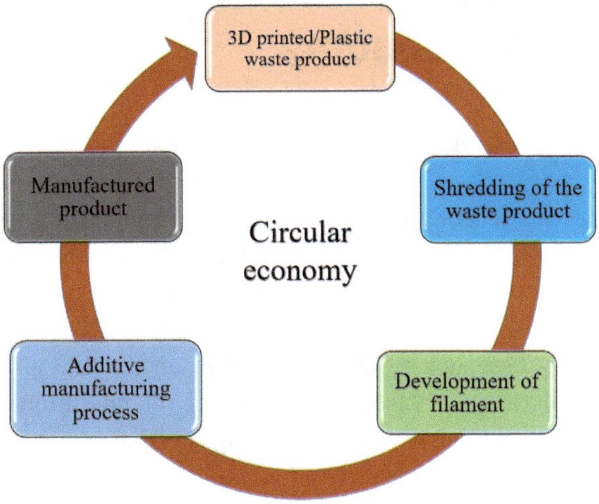

Keywords: Circular economy, plastic recycling, 3D printing, smart materials, waste management

11.1 Introduction

A circular economy introduces new ideas for the consumption, usage, and manufacturing of products that rely on the circular flow of resources (Giurco et al. 2014). AM is widely considered to be a pillar of new digital manufacturing. The circular economy, which seeks to dominate resource use while increasing business prospects, is also considered to have a strong role in the AM process. Development of no waste and having fewer physical resources are just two ways that AM can help accomplish the circular economy's goal (Kravchenko et al. 2020). The use of 3D printing is extremely flexible. The waste of 3D-printed products is a concern due to its effects on the circular economy. Economic growth improves with the process of circular economy that gives significant attention to the 3Rs of "reduce, reuse, and recycle" as well as the essential characteristics of less consumption, less emission, and higher efficiency (Akhimien et al. 2021). The linear economy model's basic structure is shown in Fig. 11.1.

"Recover, redesign, and remanufacture" procedures were added to the 3R principles, resulting in the 6R principle (Jawahir and Bradley, 2016). The circular economy model of material utilization is shown in Fig. 11.2. AM has been suggested as one of the key forces behind Industry 4.0. AM is a way for manufacturing of replica of original product or structure from 3D models, in contrast to traditional subtractive manufacturing, which involves the desired shape formation by subtracting the mate-

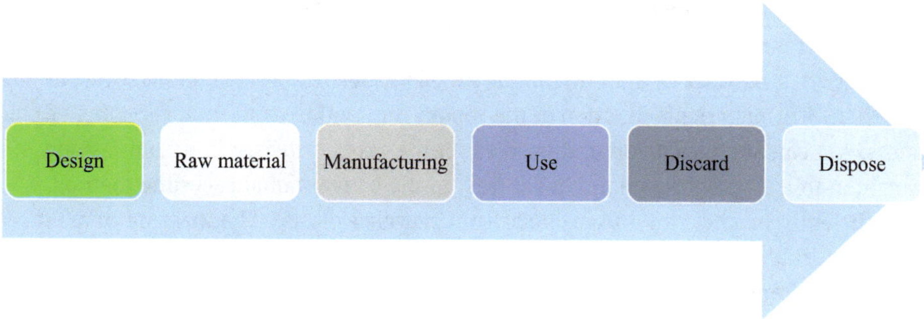

Fig. 11.1: Basic structure of the linear economy model.

rial (Gibson et al. 2021), providing the advantageous ability to develop structures with geometric complexity and material selection as opposed to traditional subtractive manufacturing, which involves shaping a product by removing material to reach a particular shape (Guo and Leu, 2013; Tofail et al. 2018).

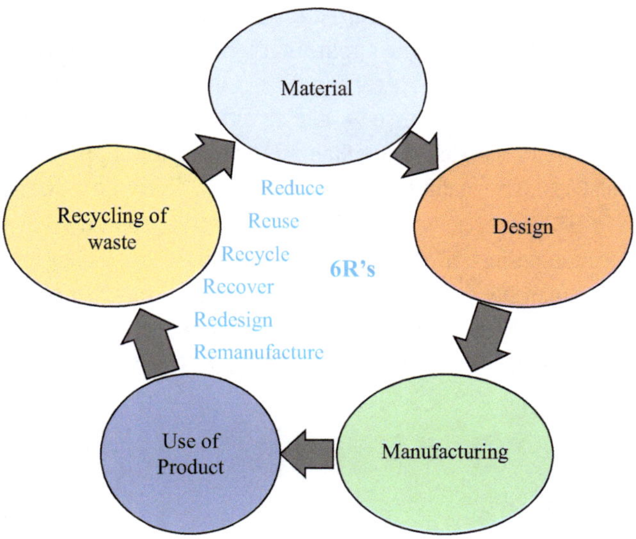

Fig. 11.2: Circular economy model of material utilization.

AM is one of the most exciting and likely to be disruptive innovations of the Industry 4.0 era, with the potential to alter the traditional manufacturing concept and promote the recognized important transition to the conception, design, and adoption of sustainable and circular business strategies (Ponis et al. 2021). The investigation of new circular economy models is necessary to lessen human impact on the environment. The internal structure of businesses and companies can significantly affect the strategies to use waste materials as raw materials or products rather than owning them (Korhonen et al.

2018; Geissdoerfer et al. 2017). Designers are essential for creating and applying various strategies within this framework. To facilitate this shift and create a framework for circular design, a new set of multidisciplinary skills is needed (Bocken et al. 2016, Sumter et al. 2021). Through the facilitation of the repair, remanufacturing, and recycling of polymers, AM can serve as a comprehensive tool that assists society in accomplishing the transition to a circular economy. In contrast to the conventional recycling method, AM can efficiently produce an end product from recycled plastic (Shanmugam et al. 2020; Sauerwein et al. 2020). The production of unnecessary stocks, typical of static and linear economic systems, could be eliminated by the ability of AM to take advantage of circular economy potentials by seeking product customization. It is more probable to occur if two essential requirements are met: consumers have the chance to build a link between the design of the item activities and the creation of an original product, the design phase is more interesting and enhanced, and the associated software tools are simplified (Angioletti et al. 2016, Syberg et al. 2021). The plastic materials consist of seven categories as per the Society of Plastic Industry: polyethylene terephthalate (PET), polystyrene (PS), high-density polyethylene (HDPE), polyvinyl chloride (PVC), low-density polyethylene (LDPE), polypropylene (PP), and other plastics (PLA, ABS, nylon, polycarbonates) (Wu et al. 2022). These materials can be recycled and used for the manufacturing of 3D-printed parts for various industrial and engineering applications. The recycled PLA composites have been manufactured with 10% and 20% carbon fiber (CF) reinforcement. The recycled PLA with 20% reinforcement of CF possessed a higher ultimate tensile strength of 47.82 MPa. The incorporation of CF in recycled PLA enhanced the mechanical properties of the composite (Al Zahmi et al. 2022). The medium density fibers (MDF) have been incorporated by weight percentage in HDPE recycle waste and the mechanical properties have been investigated for MDF/HDPE composites. HDPE with 40% MDF possessed a higher tensile strength of 30 MPa which is much better as compared to pure HDPE recycled waste (Jayaraman and Bhattacharyya, 2004). The recycled plastic composites can be used for biomedical, sensing (Singh et al. 2022), and construction applications (Conroy et al. 2006).

11.2 Background

The data available on the website www.webofknowledge.com has been used for checking the background of the study. There are just 177 results observed on the keywords "3D printing" and "circular economy" in the previous 10 years (2013–2023). The research work related to 3D printing of recycled waste material is increasing significantly and aims to reduce waste and protect the environment. The maximum number of research papers in the count was reported in 2022 as shown in Fig. 11.3.

The gaps in the previously reported literature were evaluated by analyzing the Web of Science database with the VosViewer software. A total of 5,333 terms were discovered among 177 reported publications, and 137 were filtered out after keeping

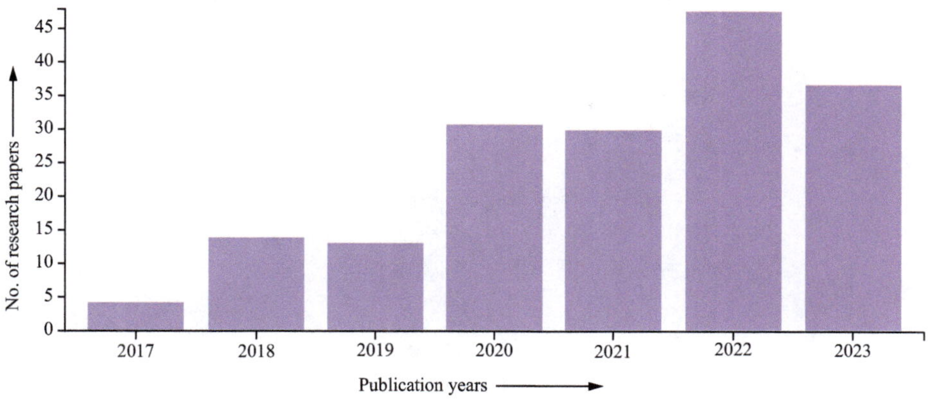

Fig. 11.3: Count of reported research papers year-wise for 3D printing of recycled waste material (circular economy) (credit: www.webofknowledge.com).

seven occurrences of terms. Again, the relationship diagram has been prepared using the 23 most relevant terms. The abundance of the terminology used in the reported literature is depicted in Fig. 11.4, and it has been proposed that earlier research projects are connected to the 3D printing of polymer materials and the manufacture of filament from raw materials. Future research includes a wide range of potential applications. As illustrated in Fig. 11.5, more investigation may be performed to make 3D-printed products from recycled waste materials and use the finished products for industrial and other engineering applications. This chapter aims to provide information related to recycling processes, number of repetitions for recycling of plastic waste, contamination level, and sustainability aspects of the circular economy in AM.

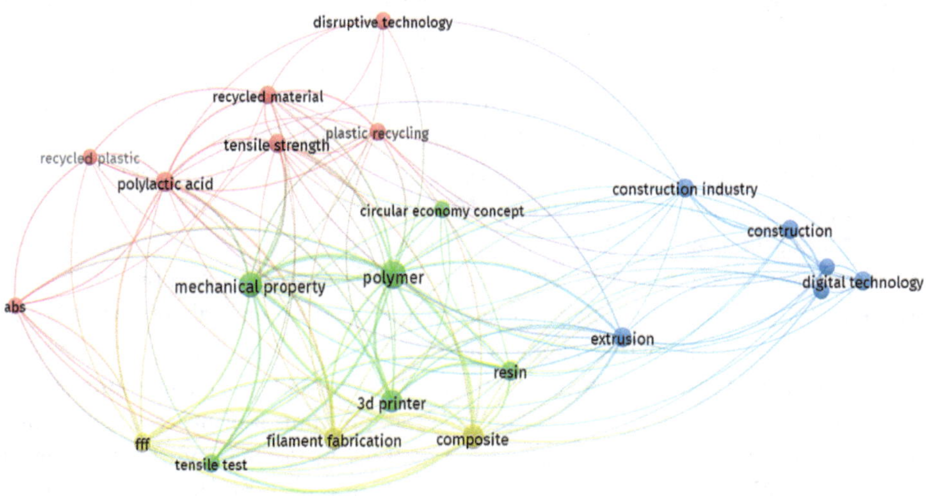

Fig. 11.4: Terms relationship in reported literature on a circular economy concept in 3D printing.

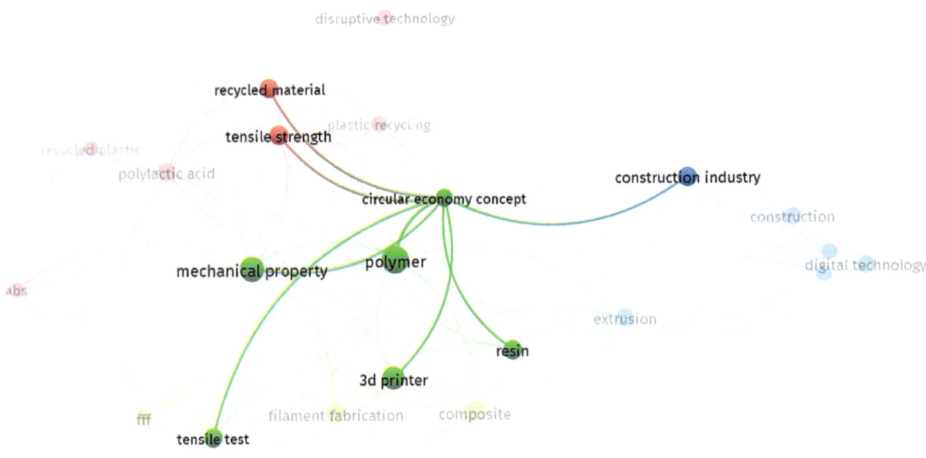

Fig. 11.5: Gaps in the previously reported literature related to circular economy and 3D printing.

11.3 3D printing of recycled materials

A new route to a circular economy employing dispersed recycling and production is made possible by the rapid technological advancement of AM. The distributed recycling via AM idea refers to the use of materials that are recycled through mechanical recycling procedures within the 3D-printing production chain (Sanchez et al. 2020). The circular economy of polymer materials has a direct impact on waste management which benefits the environmental problems. AM can enhance effectiveness in operation, performance, and resilience in use, and it can also enable reuse, repair, and recycling at the final stage of a product's beneficial lifespan. The fused deposition modeling has been used for the 3D printing of the composite structures from the polymer matrix materials. Figure 11.6(a,b) shows the illustrative diagram for the fabrication process of feedstock filament and 3D parts, respectively (Khatri et al. 2018).

Mechanical recycling, chemical recycling, and burning are the three main methods used to recycle plastics (Khatri et al. 2018). The polymer is purified from impurities in the mechanical recycling process, crushed into granulates, and then extruded to create the desired part. Incinerating waste is typically not an environmentally conscious recycling technique. It helps to recover the chemical energy entrapped in the plastics as thermal energy (Bartolome et al. 2012). The recycling of polymers from the final products has been performed which may be responsible for changes in morphological, mechanical, and structural properties. So, the control of processing parameters for recycling waste material affects the material properties and the mechanical properties can be enhanced. Figure 11.7 shows the approach for the recycling of the material or plastic waste.

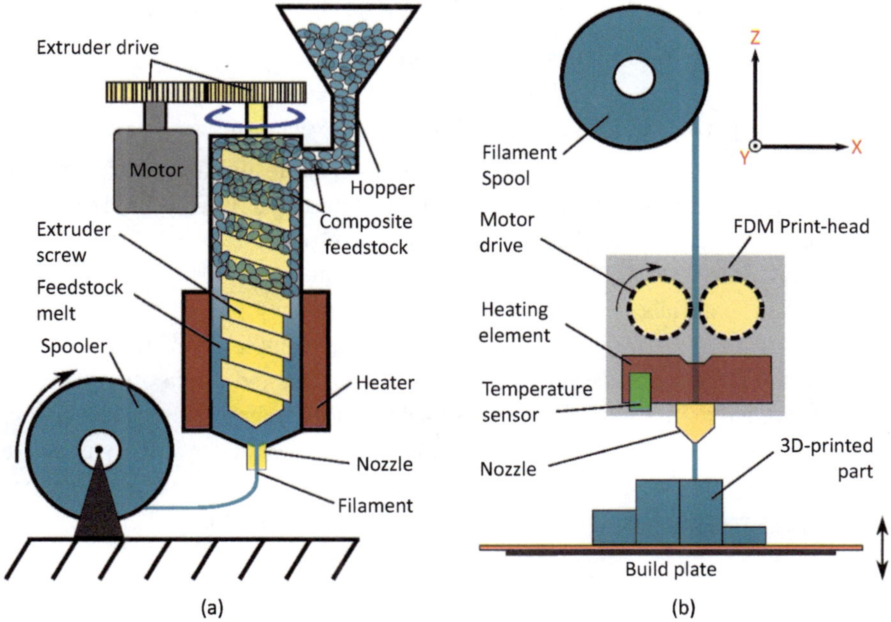

Fig. 11.6: Schematic diagram for (a) process for development of feedstock filament; (b) FDM technique for manufacturing of 3D structure (Khatri et al. 2018).

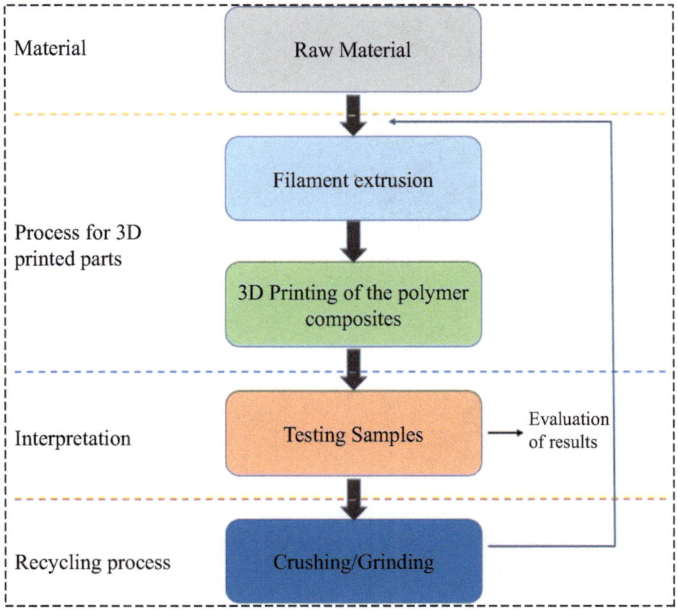

Fig. 11.7: Approach for recycling of the material.

11.4 Recycling processes to boost the circular economy of plastic waste

The plastic recycled can be of seven different forms: PET, LDPE, HDPE PP, PVC, PS, and others, most often PC and ABS. The polymers utilized in substance extrusion and AM are included in the "other" category, which includes ABS, PETG, PLA, PC, nylon, and reinforced materials. All plastics cannot be recycled as simply. In 2015, the recycled percentage of PET, HDPE, and PP was approximately 18%, 10%, and 6%, respectively. Recycled materials included LDPE, linear low-density polyethylene (LLDPE), and less than 1% of PP (Thiounn and Smith, 2020). The complexity of recycling processes ranges from practical issues with collection, sorting, and preparation (such as the presence of pollutants, adhesives, or residues) to more technical issues like chemical reactivity (Rahimi and Garcia, 2017). To improve the circular economy of waste plastic products, there are four recycling processes – primary, secondary, tertiary, and quaternary recycling as shown in Fig. 11.8 (Thiounn and Smith, 2020).

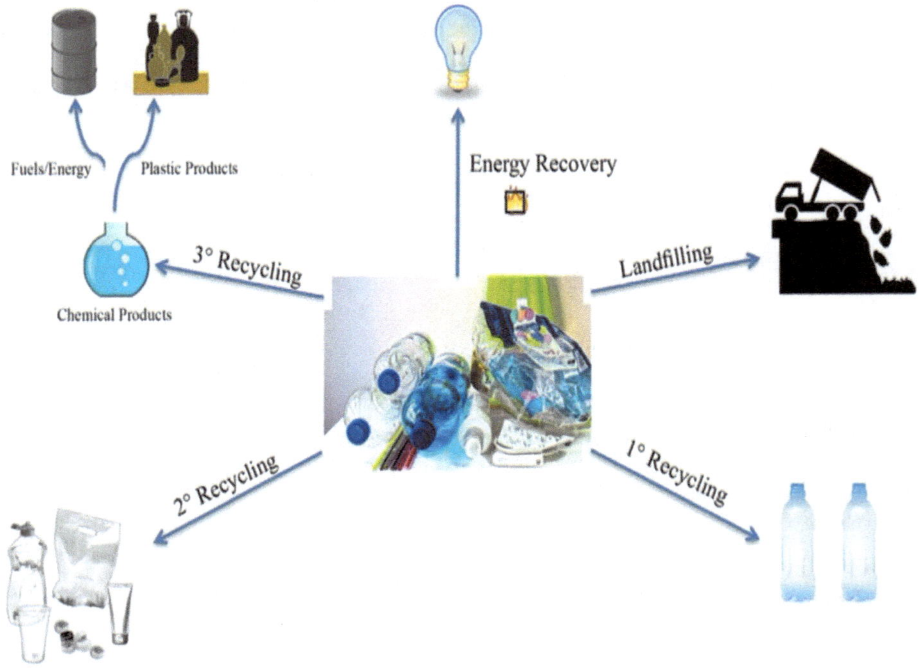

Fig. 11.8: Types of recycling processes for plastic wastes (Thiounn and Smith, 2020).

11.4.1 Primary (1°) recycling

1° recycling is also known by the name "closed-loop recycling." Primary recycling generally refers to mechanical recycling and includes the process of developing a new product from uncontaminated waste plastic without any qualities being lost (Al-Salem et al. 2009).

11.4.2 Secondary (2°) recycling

The term "2° recycling" refers to mechanical recycling, which does not alter the chemical makeup of the polymer but does physically reprocess it in some way. As a result, the polymer frequently serves purposes for which it was not intended (Ignatyev et al. 2014). This can be demonstrated by using the rubber crumb from the waste tyres as a filler in rubber flooring or park seats. The degree to which the polymers degrade when they are repeatedly reprocessed will have a substantial impact on the mechanical properties of the postrecycled product.

11.4.3 Tertiary recycling

Chemical procedures are used in tertiary recycling, also known as chemical recycling, to transform the polymer into products with value addition. Pyrolysis is a common method for processing waste polymers (Al-Salem et al. 2009) and hydrolysis (Karthikeyan et al. 2012). The finished product is then utilized as a raw material for producing fuels and polymers.

11.4.4 Quaternary (4°) recycling

The final stage of recycling involves the combustion of the polymer components, in which case the energy released is converted back into heat and this process is known as quaternary recycling. Numerous plastics that are burned also produce harmful byproducts and impurities that have a significant impact on hazardous waste management and collection (Rahimi and Garcia, 2017).

11.5 Polymer materials recycled to boost the circular economy

These are only a few instances of recyclable polymer products that can be used to encourage a circular economy. The variety of recyclable polymers keeps on increasing and processes for recycling are becoming more effective and efficient because of advancements in recycling technology. The polymers which can be recycled and utilized for manufacturing various new products are given below (Singh and Demirsoz, 2022):

11.5.1 PET

PET is recycled by the mechanical recycling process. PET is shredded in a mechanical reprocessing factory and produced in granule form. PET is a popular polymer that is used in textiles, food packaging, and beverage bottles. It is extremely recyclable and may be used to make new PET bottles and other items, such as polyester fibers for textiles and carpeting.

11.5.2 HDPE

HDPE is generally considered the easiest polymer to recycle. The HDPE has been separated (by density) and then crushed into small granules and melted down for the formation of the refined polymer. It may be easily recycled and used to create new HDPE products including bottles, containers, and more.

11.5.3 Polypropylene (PP)

Recycled PP composite filaments can be utilized in distributed manufacturing, which produces 3D-printed components with easily accessible resources. It has a high rate of recycling and can be used to make several different items, such as materials for packaging, garden furniture, and automobile parts.

11.5.4 Polystyrene (PS)

PS can be used in insulation materials, foam packing, and single-use food containers. The recycled PS materials used for applications include insulated panels, frames for pictures, and plastic molding.

11.5.5 Polyvinyl chloride (PVC)

PVC is included in pipes, wires, flooring, and frames for windows. Due to its chemical composition, PVC recycling may be more challenging, although it is still feasible with specialized techniques. PVC that has been recycled can be used to make other goods packaging, flooring, and pipes.

11.5.6 Polyethylene (PE)

PE film, a material frequently used in bags made of plastic and container films, can be reprocessed to develop new films or converted into composite timber for outdoor usage.

11.5.7 Nylon (polyamide 6)

Nylon material is present in engineered plastics, carpets, and textiles. It can be recycled to create brand-new nylon fibers for use in a variety of products, like clothes, carpeting, and automobile components.

Table 11.1 shows the previous literature related to mechanical properties observed from the composite structure in which the matrix material is recycled.

Tab. 11.1: Mechanical properties and applications of recycled polymer or plastic materials.

S. no.	Recycled material	Reinforcement	Tensile strength (MPa)	Applications	References
1.	PP	Nylon fiber	41.00	Automotive and packaging	Abraham and George (2009)
2.	PLA (10 wt.%)	PLA	37.82	Medical fields and packaging	Babagowda et al. (2018) and Cisneros-Lopez et al. (2020)
3.	PP Fiber	–	284.10	Concrete slabs	Yin et al. (2016) and Vu et al. (2020)
4.	Polyamide 12	–	33.80	Packaging, textile industry, diesel fuel lines	Vidakis et al. (2021)
5.	Acrylonitrile butadiene styrene (ABS)	–	32.00	Containers, insulation components	Vidakis et al. (2020)
6.	PLA	CF (5%)	48.00	Sports industry, automotive parts, aviation	Al-Zahmi et al. (2022)
7.	HDPE	HDPE (50 wt.%)	12.27	Composite industry, household applications	Singh et al. (2014)

11.6 Number of repetitions for recycling of waste 3D-printed plastic product

Several variables, including the specific kind of plastic, the resolution of the printed product, the chosen recycling method, and the planned use of the recycled material, might be affected by the number of repetitions of waste 3D-printed plastic recycled. Typically, the waste plastic from 3D printing is recycled first by shredding or grinding it into small grains, then melting, and converting it for the manufacture of new filament or other types of feedstocks for the manufacturing process. Due to elements like heat degradation, contamination, or modifications in the composition of the polymer, the overall quality and characteristics of the recovered material might decline with each recycling cycle (Thiounn and Smith, 2020; Vidakis et al. 2020; Vidakis et al. 2021). Depending on the type of plastic, the number of reprocessing cycles can change substantially. Some polymers, such as HDPE and PET, tend to maintain their qualities rather well over numerous recycling processes. These materials are useful for AM applications because they may be recycled numerous times without suffering substantial degradation.

11.7 Contamination level in recycling processes of 3D-printed plastic products

Primary and secondary recycling contamination can be done for numerous reasons. There may be impurities present throughout the initial manufacturing procedure, such as processing additives. The specifications of recycled materials may be impacted by these pollutants. Similar to primary recycling, secondary recycling may also contain impurities and pollutants due to insufficient removal of remaining components or decomposition intermediates from the previous printing process. Residues of cleaning solutions, support materials, or other foreign elements can be among them. Depending on the particular needs and applications of the recycled plastic, the appropriate contamination level in primary and secondary recycling varied. Higher degrees of contamination might be permissible for certain uses, especially if the recycled plastic is utilized for lower-end goods or non-food contact items. Plastic waste material that has already been recycled once before is referred to as secondary recycling. Quaternary recycling goes above and beyond tertiary recycling by incorporating extra recycling cycles. The contamination and plastic material deterioration increases as the number of recycling cycles increases (Castro et al. 2004; Flizikowski et al. 2021).

11.8 3D-printed plastic waste products in the circular economy

11.8.1 Biodegradation of 3D-printed products or materials

Metals, ceramics, and synthetic as well as natural polymers are the most common biomaterials utilized for 3D scaffolds. The most often used biomaterials in the medical sector, including tissue engineering and regeneration of bone, are biodegradable polymers like PLA. The crystallinity and molecular weight of the composition affect the biodegradation process and rate of such materials. The burning of plastic generates significant pollution, so biodegradability is a vital factor in the process of circular economy and one option to get rid of plastic is in a landfill. The products of CO_2 and H_2O's biodegradation have essentially minimal effect on the environment. The circular economy's ultimate objective is to minimize the economic activity that influences the environment. Another method of experimenting with the evolution of the circular economy is the safe management of waste (Zhu et al. 2021).

11.8.2 Boost in the circular economy of additively manufactured products by catalyst/solvents

Plastic can only be recycled once before it degrades. However, with recent advances in science and technology, the creation of catalysts and solvents may help 3D-printed goods age more quickly. The polymer can be segregated into pure polymer; however, this sort of biodegradable plastic applies to ester polymers only. Technology advancements have allowed mankind to travel to space. As a result, various 3D-printing techniques are required; however, it is unclear whether to reuse these materials. Recently, the Chinese Academy of Science revealed the use of composite polylactide/lunar regolith stimulant (PLA/CLRS-1) composites as materials in space. The mechanical properties of the composite show just a slight loss throughout the early material performance. Additionally, the dissolution of solvent approach has been applied for material reusability to entirely separate the PLA from CLRS-1, as shown in Fig. 11.9 (Li et al. 2020).

11.9 Sustainability aspects of recycled materials for AM

Engineering plastics compensate for the shortage of general-use plastics by having more significant mechanical properties and excellent stability. Engineering plastics can be challenging to recycle because of their enhanced mechanical properties and high re-

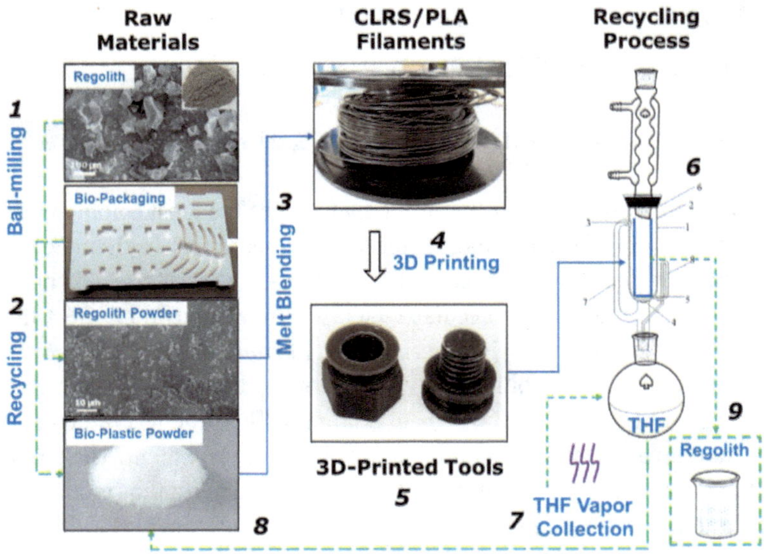

Fig. 11.9: Steps in recycling process of the components (Li et al. 2020).

liability. Reusing the benefits of engineering plastics is the ultimate objective of recycling. The durability and efficacy of operational procedures and finished products are influenced by material selection and sourcing, which are crucial steps in ensuring a more efficient and environmentally friendly recycling process. These materials include those materials that are renewable, recyclable, biodegradable, and reused. Because new materials and other inputs are needed to restore a product's value, the source of resources and the choice of materials are crucial not just for initial production but also for maintenance and repair, remodeling, and remanufacturing. One of the key benefits of AM is that it can promote local recycling of waste products, like plastic, where a plastic filament is manufactured with the aid of a recycling bot and combined with a 3D-printing machine that can use the filament directly. When attempting to recycle materials and waste, numerous sustainability considerations, such as the production of waste from the process of AM, must be taken into account in addition to the mechanical properties of materials that are recycled.

11.10 Conclusions

The majority of waste produced using 3D-printed plastics is usually disposed of in landfills or fired up, which is not a good choice because the majority of plastics decompose very slowly and their burning poses a bigger threat to the environment. Degraded plastic or recycled plastic can be used in the process of 3D printing to manufacture structures and can be used in numerous applications. At different scales and stages, AM can achieve loop closure through recycling. Although AM can increase the life cycle of 3D-printed products and close the loop, it can also open up new opportunities related to reuse, repair, and remanufacturing. The use of distributed production in 3D printing offers environmental advantages like increased efficiency of the process and less waste.

References

Abraham TN, George KE. Studies on recyclable nylon-reinforced PP composites: Effect of fiber diameter. Journal of Thermoplastic Composite Materials. 2009 Jan;22(1):5–20.

Akhimien NG, Latif E, Hou SS. Application of circular economy principles in buildings: A systematic review. Journal of Building Engineering. 2021 Jun 1;38:102041.

Al Zahmi S, Alhammadi S, ElHassan A, Ahmed W. Carbon fiber/PLA recycled composite. Polymers. 2022 May 28;14(11):2194.

Al-Salem SM, Lettieri P, Baeyens J. Recycling and recovery routes of plastic solid waste (PSW): A review. Waste Management. 2009 Oct 1;29(10):2625–43.

Angioletti CE, Sisca FG, F G F, Luglietti R, Taisch M, Rocca R. Additive Manufacturing as an opportunity for supporting sustainability through the implementation of circular economies. In Proceedings of the Summer School Francesco Turco 2016 (pp. 25–25). AIDI-Italian Association of Industrial Operations Professors.

Babagowda, Kadadevara Math RS, Goutham R, Srinivas Prasad KR. Study of effects on mechanical properties of PLA filament which is blended with recycled PLA materials. IOP conference series: Materials science and engineering. 2018 Feb 1;310:012103. IOP Publishing.

Bartolome L, Imran M, Cho BG, Al-Masry WA, Kim DH. Recent developments in the chemical recycling of PET. Material Recycling Trends and Perspectives. 2012 Mar 16;406:576–96.

Bocken NM, De Pauw I, Bakker C, Van Der Grinten B. Product design and business model strategies for a circular economy. Journal of Industrial and Production Engineering. 2016 Jul 3;33(5):308–20.

Castro MG, Remmerswaal JA, Reuter MA, Boin UJ. A thermodynamic approach to the compatibility of materials combinations for recycling. Resources, Conservation and Recycling. 2004 Dec 1;43(1):1–9.

Cisneros-López EO, Pal AK, Rodriguez AU, Wu F, Misra M, Mielewski DF, Kiziltas A, Mohanty AK. Recycled poly (lactic acid)-based 3D printed sustainable biocomposites: A comparative study with injection molding. Materials Today Sustainability. 2020 Mar 1;7:100027.

Conroy A, Halliwell S, Reynolds T. Composite recycling in the construction industry. Composites Part A: Applied Science and Manufacturing. 2006 Aug 1;37(8):1216–22.

Flizikowski J, Kruszelnicka W, Macko M. The development of efficient contaminated polymer materials shredding in recycling processes. Polymers. 2021 Feb 26;13(5):713.

Geissdoerfer M, Savaget P, Bocken NM, Hultink EJ. The circular economy – A new sustainability paradigm? Journal of Cleaner Production. 2017 Feb 1;143:757–68.

Gibson I, Rosen D, Stucker B, Khorasani M, Rosen D, Stucker B, Khorasani M. Additive Manufacturing Technologies. Cham, Switzerland, Springer, 2021.

Giurco D, Littleboy A, Boyle T, Fyfe J, White S. Circular economy: Questions for responsible minerals, additive manufacturing and recycling of metals. Resources. 2014 May 6;3(2):432–53.

Guo N, Leu MC. AM: Technology, applications, and research needs. Frontiers of Mechanical Engineering. 2013 Sep;8:215–43.

Ignatyev IA, Thielemans W, Vander Beke B. Recycling of polymers: A review. Chem Sus Chem. 2014 Jun; 7(6):1579–93.

Jawahir IS, Bradley R. Technological elements of circular economy and the principles of 6R-based closed-loop material flow in sustainable manufacturing. Procedia CIRP. 2016 Jan 1;40:103–08.

Jayaraman K, Bhattacharyya D. Mechanical performance of wood fiber–waste plastic composite materials. Resources, Conservation and Recycling. 2004 Jul 1;41(4):307–19.

Karthikeyan S, Sivakumar N, Manimekalai TK, Sathiskumar C. A review on pyrolysis of waste plastics to value-added products. Elixir Online Journal. 2012;29:8291–98.

Khatri B, Lappe K, Habedank M, Mueller T, Megnin C, Hanemann T. Fused deposition modeling of abs-barium titanate composites: A simple route towards tailored dielectric devices. Polymers. 2018 Jun 14;10(6):666.

Korhonen J, Honkasalo A, Seppälä J. Circular economy: The concept and its limitations. Ecological Economics. 2018 Jan 1;143:37–46.

Kravchenko M, Pigosso DC, McAloone TC. Circular economy enabled by additive manufacturing: Potential opportunities and key sustainability aspects. DS 101: Proceedings of Nord Design 2020, Lyngby, Denmark, 2020 Aug 12th–14th;2020:1–4.

Li H, Zhao W, Wu X, Tang H, Li Q, Tan J, Wang G. 3D printing and solvent dissolution recycling of polylactide–lunar regolith composites by material extrusion approach. Polymers. 2020 July 31;12(8):1724.

Ponis S, Aretoulaki E, Maroutas TN, Plakas G, Dimogiorgi K. A systematic literature review on additive manufacturing in the context of circular economy. Sustainability. 2021 May 26;13(11):6007.

Rahimi A, García JM. Chemical recycling of waste plastics for new materials production. Nature Reviews Chemistry. 2017 Jun 7;1(6):0046.

Rahimizadeh A, Kalman J, Fayazbakhsh K, Lessard L. Recycling of fiberglass wind turbine blades into reinforced filaments for use in additive manufacturing. Composites Part B: Engineering. 2019 Oct 15;175:107101.

Sanchez FA, Boudaoud H, Camargo M, Pearce JM. Plastic recycling in additive manufacturing: A systematic literature review and opportunities for the circular economy. Journal of Cleaner Production. 2020 Aug 10;264:121602.

Sauerwein M, Zlopasa J, Doubrovski Z, Bakker C, Balkenende R. Reprintable paste-based materials for additive manufacturing in a circular economy. Sustainability. 2020 Sep 29;12(19):8032.

Shanmugam V, Das O, Neisiany RE, Babu K, Singh S, Hedenqvist MS, Berto F, Ramakrishna S. Polymer recycling in additive manufacturing: An opportunity for the circular economy. Materials Circular Economy. 2020 Dec;2:1–1.

Singh AK, Patil B, Hoffmann N, Saltonstall B, Doddamani M, Gupta N. Additive manufacturing of syntactic foams: Part 1: Development, properties, and recycling potential of filaments. JOM. 2018 Mar;70:303–09.

Singh N, Demirsöz R. Recycling of traditional plastics: PP, PS, PVC, PET, HDPE, and LDPE, and their blends and composites. In: Nanomaterials in Manufacturing Processes. CRC Press, 2022 Aug 2, 235–58.

Singh R, Barwar A, Kumar A. Investigations on primary and secondary recycling of pla and its composite for biomedical and sensing applications. Journal of the Institution of Engineers (India): Series C. 2022 Aug;103(4):821–36.

Singh S, Deepak D, Aggarwal L, Gupta VK. Tensile and flexural behavior of hemp fiber reinforced virgin-recycled HDPE matrix composites. Procedia Materials Science. 2014 Jan 1;6:1696–702.

Syberg K, Nielsen MB, Clausen LP, van Calster G, van Wezel A, Rochman C, Koelmans AA, Cronin R, Pahl S, Hansen SF. Regulation of plastic from a circular economy perspective. Current Opinion in Green and Sustainable Chemistry. 2021 Jun 1;29:100462.

Sumter D, de Koning J, Bakker C, Balkenende R. Key competencies for design in a circular economy: Exploring gaps in design knowledge and skills for a circular economy. Sustainability. 2021 Jan 14;13(2):776.

Thiounn T, Smith RC. Advances and approaches for chemical recycling of plastic waste. Journal of Polymer Science. 2020 May 15;58(10):1347–64.

Tofail SA, Koumoulos EP, Bandyopadhyay A, Bose S, O'Donoghue L, Charitidis C. AM: Scientific and technological challenges, market uptake and opportunities. Materials Today. 2018 Jan 1;21(1):22–37.
Vidakis N, Petousis M, Maniadi A, Koudoumas E, Vairis A, Kechagias J. Sustainable additive manufacturing: Mechanical response of acrylonitrile-butadiene-styrene over multiple recycling processes. Sustainability. 2020 Apr 27;12(9):3568.
Vidakis N, Petousis M, Tzounis L, Maniadi A, Velidakis E, Mountakis N, Kechagias JD. Sustainable additive manufacturing: Mechanical response of polyamide 12 over multiple recycling processes. Materials. 2021 Jan 19;14(2):466.
Vu HH, Do TG, Nguyen TT. The application of polypropylene fiber for reinforced concrete beams and slabs. IOP Conference Series: Materials Science and Engineering 2020 Jun 1;869(7):072042. IOP Publishing.
Wu H, Mehrabi H, Karagiannidis P, Naveed N. Additive manufacturing of recycled plastics: Strategies towards a more sustainable future. Journal of Cleaner Production. 2022 Feb 10;335:130236.
Yin S, Tuladhar R, Riella J, Chung D, Collister T, Combe M, Sivakugan N. Comparative evaluation of virgin and recycled polypropylene fiber reinforced concrete. Construction and Building Materials. 2016 Jul 1;114:134–41.
Zhu C, Li T, Mohideen MM, Hu P, Gupta R, Ramakrishna S, Liu Y. Realization of circular economy of 3D printed plastics: A review. Polymers. 2021 Feb 27;13(5):744.

Kanwer Ajit Singh*, Amrinder Pal Singh, Rupinder Singh, Yang Wei

12 Life cycle analysis for economic and environmental justification of 3D-printed smart functional prototypes

Abstract: Food-grade thermoplastic polyethylene terephthalate glycol (PETG) is frequently used for food product packaging. According to a bibliographic review of Web of Sciences data, there is little material accessible on the use of smart PETG containers for real-time monitoring of food goods. For the circular economy, PETG thermoplastic (postconsumer waste) material has been collected, segregated, and processed mechanically without the use of any solvent in this study. By concentrating on the creation of smart containers made of PETG that can be 3D-printed, this study seeks to close the gap in the circular economy approach. To determine the remaining shelf life of particular foods, these containers can be used in conjunction with a smartphone application that utilizes Bluetooth technology. For the experimental study, a solution of mustard oil (*Brassica juncea*), containing allyl isothiocyanate, along with NaCl was used as a stimulus. One gram of NaCl and 10 mL of oil made up the solution. The resonant frequency peak changed from 2.75 to 2.70 GHz when the PETG substrate was exposed to this mixture and 5 mL of water to represent moisture. The dielectric constant (ε), which shifted from 2.67 to 2.73, was another notable change. Further voltage-resistance measurements on the ring resonator revealed that, when exposed to stimuli with and without moisture, respectively, the resistance of PETG reduced from 47.08 to 42.5. According to this study, PETG containers fitted with these sensors might be used to monitor the quality and safety of specific food products in real time.

Acknowledgment: The authors are thankful to NITTTR Chandigarh and Panjab University, Chandigarh, for laboratory support.

***Corresponding author: Kanwer Ajit Singh**, University Institute of Engineering and Technology, Panjab University, Chandigarh, India, e-mail: kanwerajit2036@gmail.com

Amrinder Pal Singh, University Institute of Engineering and Technology, Panjab University, Chandigarh, India

Rupinder singh, Department of Mechanical Engineering, National Institute of Technical Teachers Training and Research, Chandigarh, India

Yang Wei, Department of Engineering, Nottingham Trent University, Nottingham, UK

https://doi.org/10.1515/9783111081632-012

Graphical abstract:

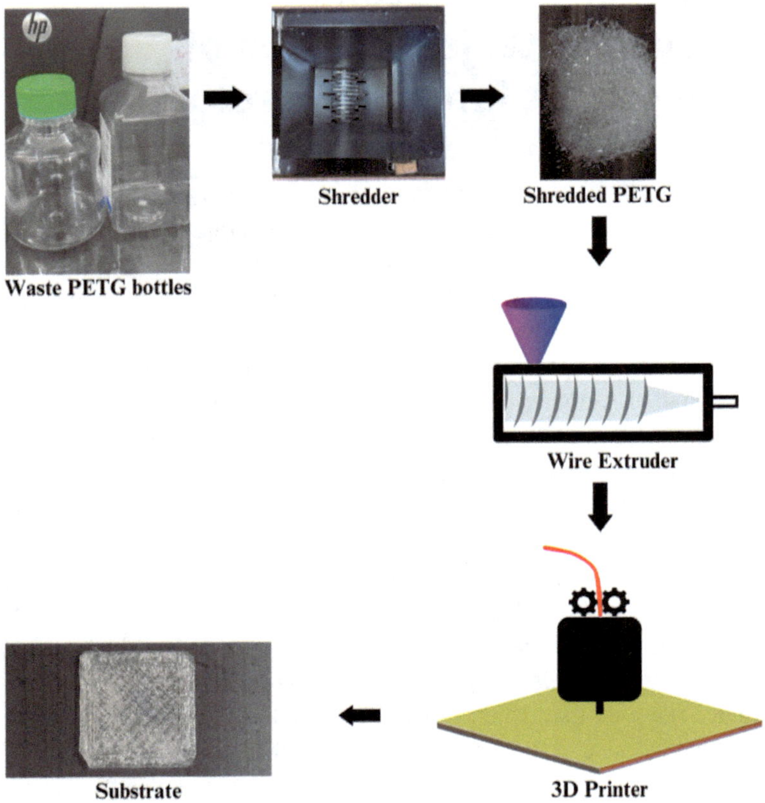

Keywords: PETG, circular economy, 3D printing, *Brassica juncea*, smart containers

12.1 Introduction

Polyethylene terephthalate glycol (PETG) is a thermoplastic and food-grade material that is preferably used in the packaging of food articles (FA). Due to the effect of globalization, there is a huge volume of packaged FA that is imported and exported across the world (Mukhopadhyay et al. 2015). These packaging materials/polymers are nonbiodegradable; hence they are piling up the waste material, which will be sustained with time. There is an urgency to eliminate or reduce nonbiodegradable waste. Recycling is one of the options because in this era it is not possible to eliminate the use of packaging material.

PETG is one of the major constituents of nonbiodegradable wastes. In the past few years, PETG has been successfully recycled and used (Latko-Durałek et al. 2019).

Mechanical, thermal, rheological, and morphological characterization of recycled PETG has been performed to validate its 3D printability. PETG is less brittle as compared to PETG because glycol retains its amorphous structure. Fused filament fabrication (FFF) is the manufacturing technique in which material is deposited layer by layer to form a product. To evaluate the shelf life of FA with a smartphone and a Bluetooth application, 3D-printed PETG has been employed as the substrate material for smart containers as an IoT solution. 3D-printed substrates have been used for the fabrication of ring resonators instead of conventional manufacturing processes. 3D printing provides a high degree of freedom in design parameters like infill density, speed, raster angle, orientation, and flow rate (Harle et al. 2017). Characteristics of ring resonators depend on the substrate material and the structure of the substrate. The design and density of the structure of the substrate change the radio frequency (RF) characterization of the ring resonator (Zhang et al. 2015). This characteristic of the ring resonator may be used to sense the stimulus. PETG-based ring resonator can be used as a sensor and its RF characteristics may be explored to detect stimuli for online health monitoring of FA.

12.2 Background

A lot of studies have been reported on 3D printing, sensors, PETG polymer, and waste management separately. However to ensure the research gap for these above-mentioned keywords, a bibliographic analysis was performed. Web of science data was used for this purpose. 3D printing was used as a keyword initially, and 38,230 results were obtained. The 1,000 most recent results were selected and analyzed with VOS viewer (open source software). A minimum of 5 times occurrence of 3D printing was selected and 804 terms met the threshold out of 20,192 terms. For the final analysis, 482 terms out of 804 were selected (as the top 60%) (Tab. 12.1). Using Tab. 12.1, a networking diagram (Fig. 12.1) was prepared for bibliographic analysis.

This same analysis was performed for the keywords 3D printing and PETG. The 91 results were obtained and these results were processed with a VOS viewer. The bibliographic analysis based on Tab. 12.2 is shown in Fig. 12.2.

Similarly, 3D printing, PETG, and sensors were used as keywords on the Web of Sciences database, and only six publications were found and these results were processed with a VOS viewer. The network diagram is shown in Fig. 12.3.

Further, 3D printing, PETG, sensors, and smart containers were used as keywords for bibliographic analysis on the Web of Science database and only one research paper was obtained. This work is the continuation of the previously reported study by Singh et al. (2023) on the use of PETG from a recycling viewpoint.

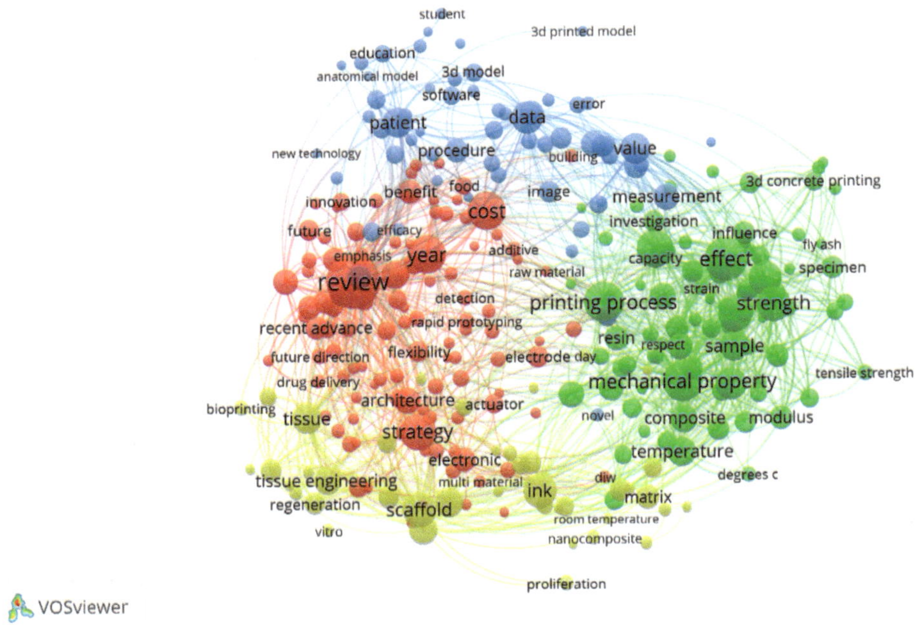

Fig. 12.1: Bibliographic analysis for keyword 3D printing.

Tab. 12.1: Terms, occurrences, and relevance scores with the 3D-printing keyword.

Id	Terms	Occurrences	Relevance scores
1	3D concrete printing	32	2.5286
2	3D-printed model	12	2.3469
3	3D-printed part	15	1.1464
4	3D-printed product	10	1.0121
5	3D-printed structure	12	0.5582
6	3D-printing application	16	0.8206
7	4D printing	24	1.4842
8	Actuator	33	0.8673
9	Additive manufacturing technique	14	0.9124
10	Aerospace	11	0.5045
11	Anatomical model	11	2.2478
12	Architecture	54	0.5916
13	Bioink	11	1.9362
14	Biomaterial	38	1.4086
15	Biomedical application	34	0.946
16	Biomedical device	12	1.5165
17	Bioprinting	19	1.9679
18	Bone	32	0.9631
19	Building	16	0.8384
20	CAD	11	0.7852

Tab. 12.1 (continued)

Id	Terms	Occurrences	Relevance scores
21	Cell	67	0.9206
22	Clinical application	16	1.0307
23	Complex shape	11	0.5003
24	Composite	60	0.5724
25	Compression	17	1.1014
26	Concrete	36	1.7838
27	Construction industry	15	1.1637
28	Conventional method	13	0.6963
29	Current challenge	13	1.949
30	Deformation	29	0.8717
31	Degrees c	22	1.4084
32	Density	44	0.4763
33	Deposition modeling	16	0.6389
34	Design methodology approach	24	1.3747
35	Diameter	33	0.8869
36	Digital light processing	25	0.9529
37	Dimensional printing	22	1.0413
38	Direct ink writing	23	1.1295
39	Disadvantage	16	0.7513
40	Drug	29	1.027
41	Drug delivery	18	1.431
42	Effect	134	0.6408
43	Electrode	39	0.4412
44	Electron microscopy	16	1.1453
45	Electronic	48	0.9917
46	Emergence	12	0.5411
47	Emphasis	16	1.7199
48	Evaluation	39	0.5662
49	Evidence	12	1.4358
50	Experiment	56	0.4539
51	Experimental result	32	0.7293
52	Failure	19	2.1045
53	FDM	37	0.3953
54	FFF	16	1.6266
55	Fiber	23	0.9764
56	Filament	87	0.9164
57	Filament fabrication	20	1.426
58	Flexibility	37	0.7589
59	Fly ash	12	2.3445
60	Food industry	12	2.1088
61	Formulation	54	0.4371
62	Functional material	20	1.3105
63	High performance	11	1.1889
64	High resolution	23	0.7501
65	Inkjet	10	0.6267

Tab. 12.1 (continued)

Id	Terms	Occurrences	Relevance scores
66	Inkjet printing	17	0.4817
67	Innovation	27	1.0786
68	Material extrusion	12	0.2774
69	Mechanical performance	15	1.626
70	Mechanical property	117	0.4946
71	Mechanical strength	21	0.4778
72	Medical device	25	0.8863
73	Mixture	28	2.024
74	Multimaterial	15	1.0686
75	Operation	23	0.6125
76	Opportunity	60	0.5889
77	Originality value	25	1.3146
78	Past decade	14	1.1165
79	Plastic	18	0.463
80	Polymerization	22	0.5189
81	Porosity	29	0.3696
82	Porous structure	10	0.8993
83	Practical application	10	0.7271
84	Preoperative planning	11	2.7398
85	Presence	14	0.6618
86	Printability	51	0.3302
87	Printable material	15	0.8238
88	Printed material	14	0.6466
89	Printed object	17	0.5985
90	Printing parameter	40	0.4327
91	Printing process	117	0.1564
92	Prosthesis	15	1.1158
93	Rapid prototyping	23	0.4216
94	Reinforcement	19	1.7248
95	Revolution	14	1.169
96	Scaffold	86	0.8028
97	Selective laser sintering	15	0.6142
98	SEM	17	1.1089
99	Simulation	56	0.4284
100	Stiffness	21	0.9593
101	Surface roughness	16	0.7429
102	Surgery	33	1.9176
103	Sustainability	20	1.0545
104	Systematic review	10	2.2121
105	Temperature	66	0.6604
106	Tensile strength	22	2.1553
107	Tissue	71	1.1092
108	Tissue engineering	58	1.1056
109	Tomography	26	1.0979
110	Traditional method	14	0.8072

Tab. 12.1 (continued)

Id	Terms	Occurrences	Relevance scores
111	Versatility	18	0.8638
112	Viscosity	41	0.8833
113	Vitro	14	1.5269
114	X-ray diffraction	10	1.3836

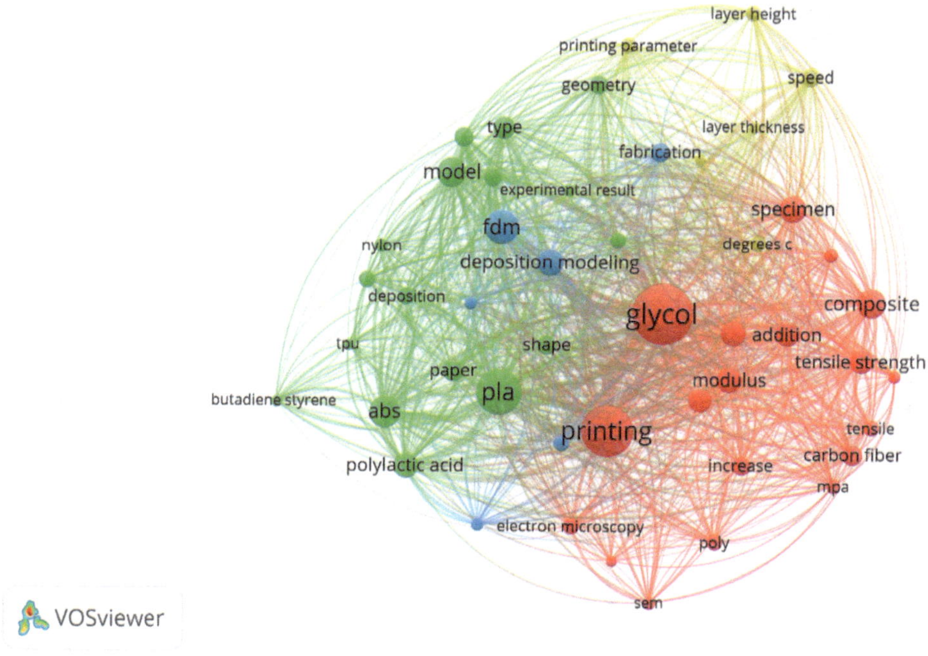

Fig. 12.2: Bibliographic analysis of keyword 3D printing and PETG.

Tab. 12.2: Terms, occurrences, and relevance score with the 3D printing and PETG keyword.

Id	Terms	Occurrences	Relevance scores
1	3D printer	16	0.333
2	ABS	43	1.0952
3	Acrylonitrile butadiene styrene	12	0.7762
4	Addition	30	0.5241
5	Butadiene styrene	10	2.7723
6	Carbon fiber	24	1.0225
7	Composite	38	1.135
8	Degrees c	16	0.3971

Tab. 12.2 (continued)

Id	Terms	Occurrences	Relevance scores
9	Deposition	17	1.1563
10	Deposition modeling	34	1.1817
11	Differential scanning calorimetry	10	0.9048
12	Electron microscopy	17	0.8437
13	Experiment	21	0.5288
14	Experimental result	11	0.6924
15	Fabrication	21	0.5368
16	FDM	47	1.142
17	FFF	28	1.3858
18	Filament fabrication	31	1.3621
19	Flexural strength	12	2.2628
20	Geometry	20	0.5988
21	Glycol	111	0.2689
22	Hand	10	0.7182
23	Increase	21	0.7764
24	Layer height	14	2.3206
25	Layer thickness	10	1.2235
26	Low cost	11	0.8549
27	Mechanical performance	14	1.4455
28	Model	38	0.6686
29	Modulus	32	0.4018
30	MPa	14	1.6658
31	Nylon	13	1.6737
32	Paper	24	0.4811
33	PLA	73	0.6221
34	Plastic	17	1.1004
35	Poly	16	0.4825
36	Polyethylene terephthalate	18	0.8657
37	Polylactic acid	32	1.1091
38	Printing	87	0.2183
39	Printing parameter	14	1.3464
40	SEM	12	1.0137
41	Shape	22	0.3678
42	Specimen	34	0.577
43	Speed	19	1.7211
44	Tensile	16	1.0317
45	Tensile strength	29	0.9673
46	Term	16	0.435
47	TPU	10	2.2054
48	Type	26	0.7863

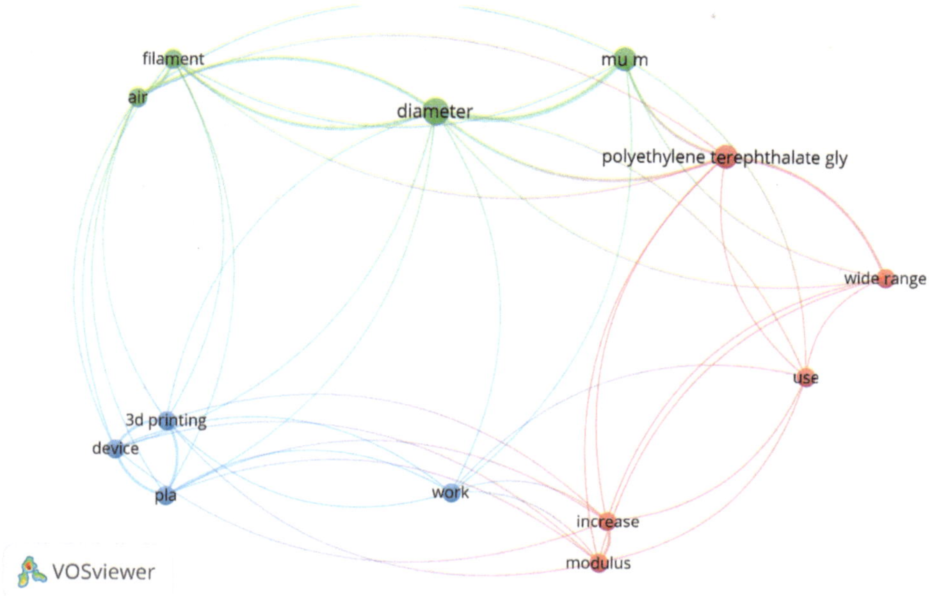

Fig. 12.3: Bibliographic analysis of keywords 3D printing, PETG, and sensor.

12.3 PETG

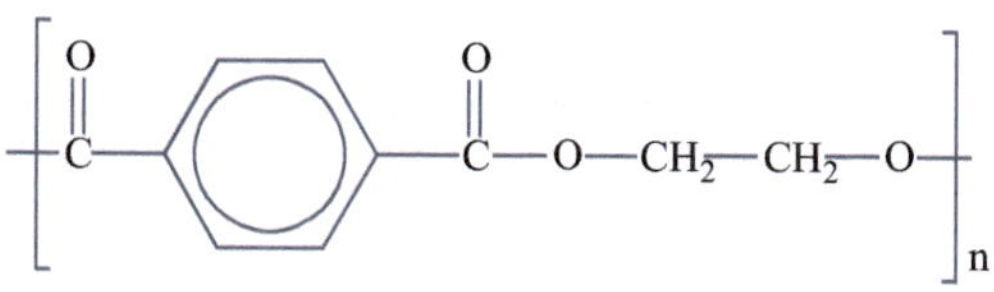

Fig. 12.4: Chemical formula of PETG.

An improved version of thermoplastic PET is PETG. PET is copolymerized with glycol to form PETG. It is more flexible, formable, and has more impact strength as compared to PET. Its less wrapping and proper layer adhesion make it a more suitable material for 3D printing. Figure 12.4 shows the chemical structure of the PETG. PETG has been recommended for the packaging of FA which is stored in a liquid atmosphere (Moreno Nieto et al. 2021). Its anticorrosion behavior, physical and chemical resistance, and optical properties (Petrov et al. 2021) make it a preferable packaging material (Paszkiewicz et al. 2017, Habel et al. 2018, Bałdowska-Witos et al. 2020). PETG is a good candidate for 3D printing because of its printability (Szykiedans et al. 2017). It has high impact resistance and ductility as compared to PET (Guessasma et al. 2019). PETG is also a biocompatible material (Balint et al. 2021, Sölmann et al. 2021) and has been used in tissue engineering. This characteristic of PETG widens its range of application in the medical field. Water bottles and packaging bags made up of polymer are piling up the waste load on Earth (Epps et al. 2021, Singh et al. 2022). These polymers are nondegradable hence their recycling is the only way to mitigate the burden of polymer wastage. The filament of recycled PETG can be used for 3D printing (Mikula et al. 2021). There is an abundance of PETG waste that can be recycled mechanically (Shamsuyeva et al. 2021) or with the help of enzymes (Tamoor et al. 2021) and used again (Huang et al. 2018).

12.4 Recycling (economic and environmental aspects)

The high demand for PETG leads to the accumulation of large amounts of waste material on the Earth. To protect the environment, a circular economy approach should be implemented to recycle and reuse PETG. Some techniques are mentioned below for the recycling and reusing of PETG

12.4.1 Mechanical recycling

PETG can be mechanically recycled, involving processes such as shredding, melting, and regranulation. During mechanical recycling, PETG is cleaned, melted, and then reformed into new PETG products, such as sheets or containers.

Fabrication of 3D printer filament: PETG can be recycled and 3D printer filament will be fabricated, allowing the reuse of waste PETG in various applications.

– Chemical recycling:

In this recycling, the polymer is broken into constituent monomers by methods like hydrolysis and glycolysis. These monomers can be further processed to fabricate PETG.

– Depolymerization:

As the name suggests, the PETG polymer is broken down into base monomers, terephthalic acid, and ethylene glycol. These monomers can be used again to produce PETG.

– Blending of recycled PETG with virgin PETG:

Recycled PETG can be blended with virgin PETG to retain the properties of PETG. This does not allow recycled PETG to lose its characteristics.

– Blending with reinforced materials:

The composite material can be fabricated by adding some reinforced material to the recycled material. Some of the reinforced materials are graphene, fiberglass, carbon nanotubes, etc.

PETG can be used in the production of composite materials by combining it with other materials like fiberglass or carbon fiber. This application can extend the life of PETG and reduce the need for virgin materials.

– Upcycling PETG:

Upcycling involves transforming PETG waste into higher-value products. For example, PETG bottles can be repurposed into textiles, construction materials, or other innovative products.

– Biological recycling of PETG:

Some research explores the use of enzymes or microorganisms to break down PETG into simpler compounds, which can be used in various applications.

– Additive-based recycling:

Innovative additives can be incorporated into PETG to improve its recycling properties. These additives can enhance the compatibility of recycled PETG with virgin PETG during processing.

In this study, postconsumer waste PETG bottles were collected and their tags were removed. Waste bottles were mechanically shredded into fine pieces. Further, shredded PETG material was poured into the hopper of the wire extrusion machine, and the 3D-printer filament of uniform diameter (1.75 mm) was produced. This fabricated filament was used to print substrate using Ender Pro 3, a 3D printer (open-source). The adopted methodology is shown in Fig. 12.5.

12.5 Smart containers

Smart containers are intelligent containers that are used to store materials such as food. These containers are equipped with sensors, data processing capabilities, and communication technology. These containers collect, monitor, and transmit information about the content and the environment of the material. Information transmitted through smart containers can be used for real-time monitoring of the products.

Smart containers possess the following properties (Kuswandi et al. 2011):

- Provide information about product authenticity.
- Provide product history or condition to the user.
- Indicate any change in the product or the package environment.
- Enhance the color, fragrance, flavor, aroma, and texture of the product.
- Prevent food spoilage and maintain integrity.
- Assist with the opening of the container.

- Location tracking.
- Security purpose.
- Management of inventory.

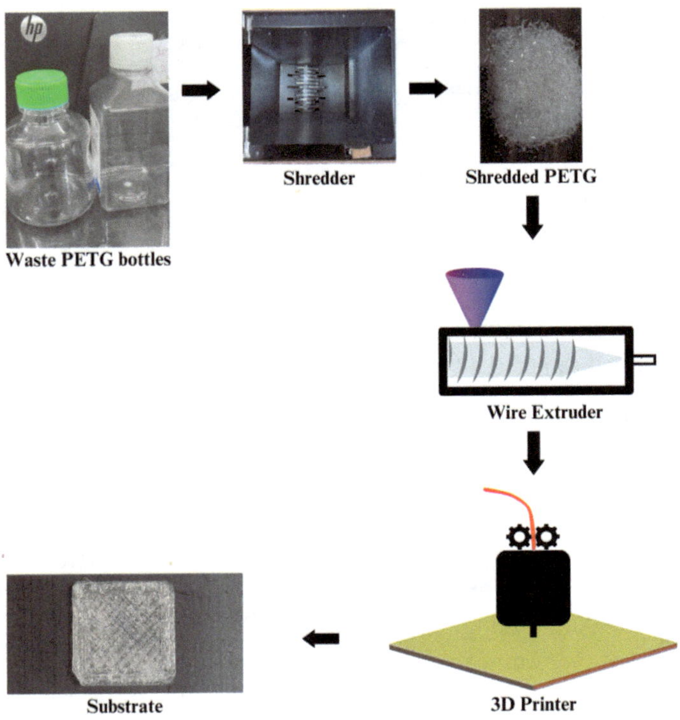

Fig. 12.5: Recycling process of PETG material.

12.6 Melt flow index

MFI is the measure of the weight of the molten material coming out of a specific-sized nozzle in 10 min under the action of the applied load. Rheological analysis of PETG polymer is performed using the Pacorr Melt Flow Index Tester (Fig. 12.6). ASTMD1238 standard is used to perform the testing (Reinaldo). A weight of 2.16 kg was used as a load to draw molten material through the nozzle of the tester. This test was performed in triplicates and the mean of those values was considered as the MFI of the PETG: 18.3 g/10 min, 15.50 g/10 min, and 16.90 g/10 min are obtained values of the MFI. The mean value of these 3 is 16.90 g/10 min considered as the MFI of PETG.

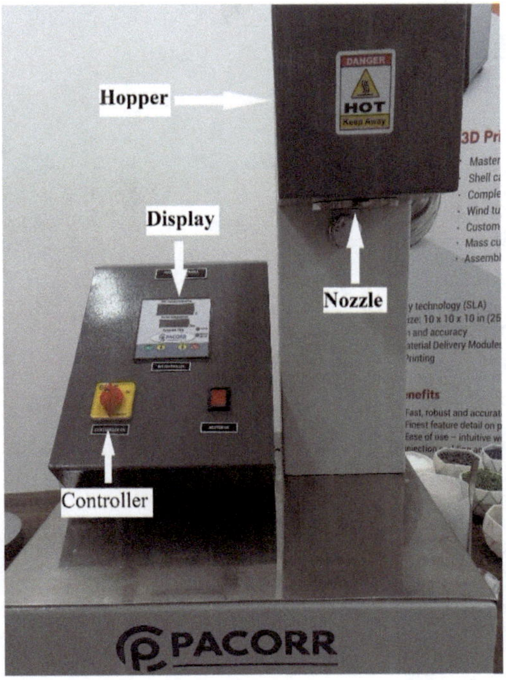

Fig. 12.6: Pacorr's melt flow index tester.

12.7 Tensile testing

Mechanical characterization of PETG was performed on the Biss UTM (Fig. 12.7). It is a multioperational testing machine for tensile, compression, bending, and torsional tests. Table 12.3 shows the specifications of the UTM used in this study.

Tab. 12.3: Specifications of UTM machine.

1	Actuator capacity	400 N
2	Torque capacity	2 Nm
3	Total actuator stroke	45 mm
4	Horizontal daylight (*C*)	74 mm
5	Vertical daylight (*B*)	270 mm max 225 mm min
6	Height (*H*)	480 mm max

Tab. 12.3 (continued)

7	Width (*W*)	160 mm
8	Depth (*D*)	250 mm
9	Frame height (*E*)	418 mm
10	Load cell height (*F*)	81.5 mm
11	Adapter height (*G*)	38 mm
12	Weight	20 kg

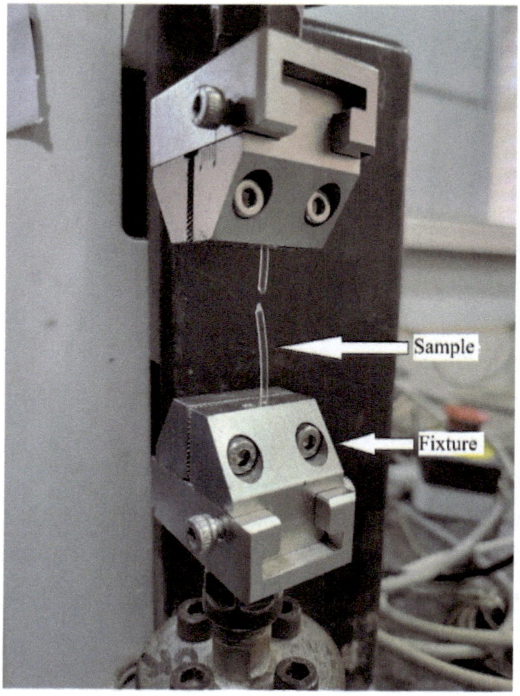

Fig. 12.7: PETG wire in the fixtures of the UTM machine.

PETG wire of diameter 1.75 mm was used to obtain the tensile strength at the rate of 10 mm/min. The obtained value of the tensile strength of the PETG is 45.28 MPa. Figure 12.8 shows the stress–strain diagram of the PETG wire.

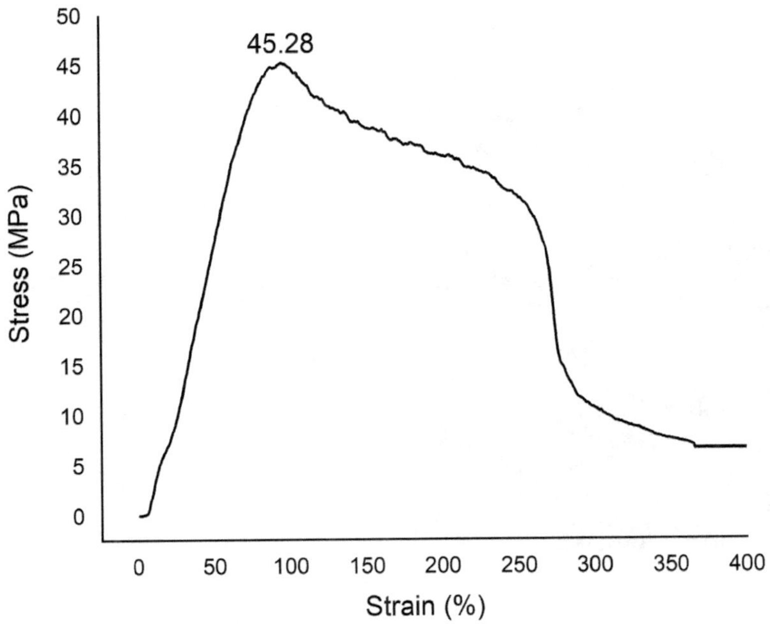

Fig. 12.8: Stress–strain diagram of PETG wire.

12.8 3D printing

3D printing has many advantages over conventional manufacturing processes and can be used for functional/nonfunctional prototypes (Fina et al. 2017, Revilla-León et al. 2019, Fan et al. 2020).

– Extrusion-based printing/FFF:

Molten thermoplastic or composite materials are drawn through the heated nozzle and deposited layer by layer on the heated bed. Movement of the bed and nozzle deposits the material in a required pattern. There can be two nozzles for the extrusion of different materials at the same time. This is generally used for polymer and bioprinting (Credi et al. 2018). Figure 12.9 shows various parts of the 3D printer used in this study.

For R-F characterization, the substrate of the ring resonator was 3D-printed. Initially, a substrate of 50 × 40 mm was designed, and the design was saved in ".STL" format. The Ultimaker Cura software was used as a slicer and the ".STL" file was converted into a G-code file. Table 12.4 shows the selected process parameters.

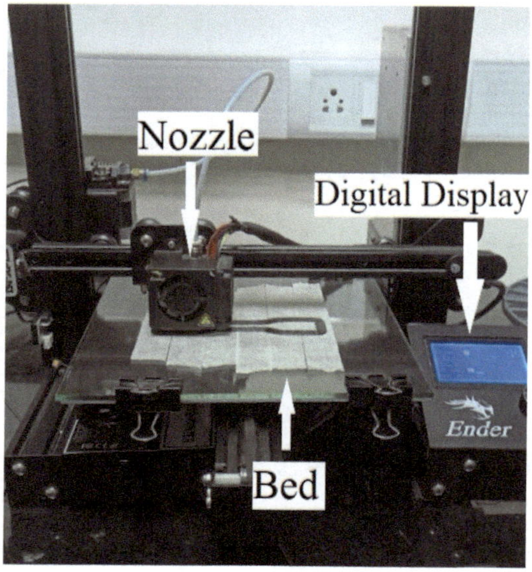

Fig. 12.9: Various parts of 3D printer.

Tab. 12.4: Parameters of the 3D printer for PETG substrate.

S. no.	Parameter	Value
1	Nozzle temperature	240 °C
2	Speed	25 mm/min
3	Build plate temperature	75 °C
4	Flow rate	100%

12.9 Ring resonator

The 3D-printed substrate was used to fabricate the ring resonator as per Tab. 12.5 as shown in Fig. 12.10. Dimensions of the ring resonator were evaluated at the 50 Ω impedance and 2.45 GHz resonating frequency. The following equations were used to obtain the dimensions of the ring resonator (Singh et al. 2022):

$$2\pi R = n\lambda_g \qquad (\text{for } n=1,\ 2,\ 3) \tag{12.1}$$

where R is the mean radius, n the order of harmonic oscillations, then, guided wavelength (λ_g) is calculated from the following relation:

$$\lambda_g = \frac{c}{\sqrt{\varepsilon_{\text{eff.}}}} \times \frac{1}{f} \tag{12.2}$$

where c is the speed of light, f the resonating frequency, and ε_{eff} the effective dielectric constant.

Then, the effective dielectric constant is given by

$$\varepsilon_{eff.} = \left[\frac{\varepsilon_r + 1}{2}\right] + \frac{1}{\left[\left(\frac{\varepsilon_r - 1}{2}\right)\left(1 + \frac{12h}{w}\right)\right]} \tag{12.3}$$

where h is the substrate thickness, w the width of the ring, and ε_r the relative dielectric constant.

Now, feedline length

$$\text{Feedline length} = \frac{\lambda_g}{4} \tag{12.4}$$

Then,

$$\text{Inner radius} = R - \frac{w}{2} \tag{12.5}$$

and

$$\text{outer radius} = R + \frac{w}{2} \tag{12.6}$$

Tab. 12.5: Dimensions of the ring resonator.

S. no.	Parameter	Value
1	Length of substrate	50 mm
2	Width of substrate	40 mm
3	The outer radius of the ring	12.47 mm
4	The inner radius of the ring	11.07 mm
5	Mean radius	11.77 mm
6	Feed line length	11.27 mm
7	Width of the feed line	1.4 mm
8	Copper tape width	0.8 mm

12.10 Vector network analysis (VNA)

VNA setup (Fig. 12.11) was used to obtain the S_{21} parameters of the designed ring resonator (Tab. 12.6). Figures 12.12–12.14 show the resonating frequency of various samples used in this study. For the analysis, both ports of the VNA were calibrated using a calibration kit.

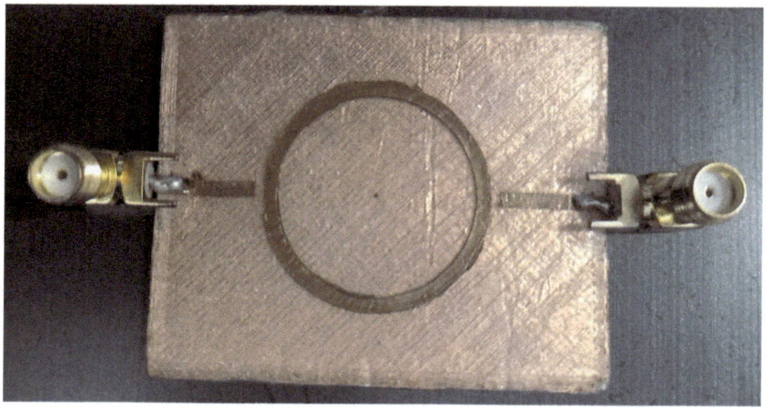

Fig. 12.10: Ring resonator of PETG substrate.

Tab. 12.6: Values of insertion loss (S_{21}), dielectric constant (ε), and resonating frequency (GHz) for various samples.

S. no.	Sample	Insertion loss S_{21} (dB)	Dielectric constant (ε)	Resonating frequency (GHz)
1	PETG substrate without exposure to stimuli	−51.72	2.62	2.77
2	PETG substrate without moisture	−51.02	2.68	2.75
3	PETG substrate with moisture	−62.01	2.74	2.7

12.11 Fourier transformation infrared (FTIR) spectroscopy

Figure 12.15 shows the absorption spectrum of PETG substrate without exposure to the stimuli. Wavenumber 1,712.48 cm^{-1} represents the double bond of carbon and oxygen (C=O) that is the easter group. Absorbance peaks at 2,852.2 cm^{-1} and 2,925.48 cm^{-1} represent C—H bond. The observed absorbance peaks at these wavenumbers validate the material as PETG. Figure 12.16 represents the FTIR spectra of the PETG substrate which was exposed to stimuli (without moisture). The presence of allyl isothiocyanate (main constituent oil of mustard oil) was denoted by the peak at 1,631 cm^{-1}; further a range of wavenumber from 3,200 cm^{-1} to 3,700 cm^{-1} represents the Cl^- element of NaCl. These FTIR spectra validate the presence of mustard oil and NaCl on the PETG substrate.

Figure 12.17 shows the absorbance peak at 1612.2 cm^{-1}, representing the water molecule. This is the case when PETG substrate was exposed to the stimuli (with moisture). Other peaks represent the PETG, oil, and NaCl.

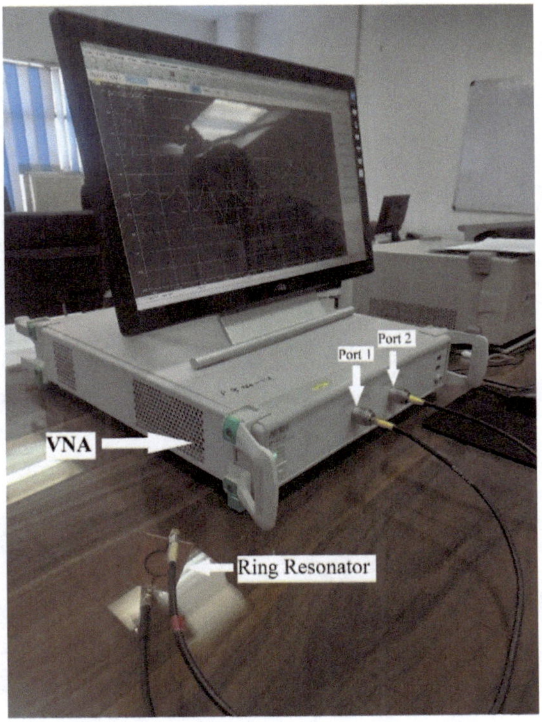

Fig. 12.11: VNA apparatus.

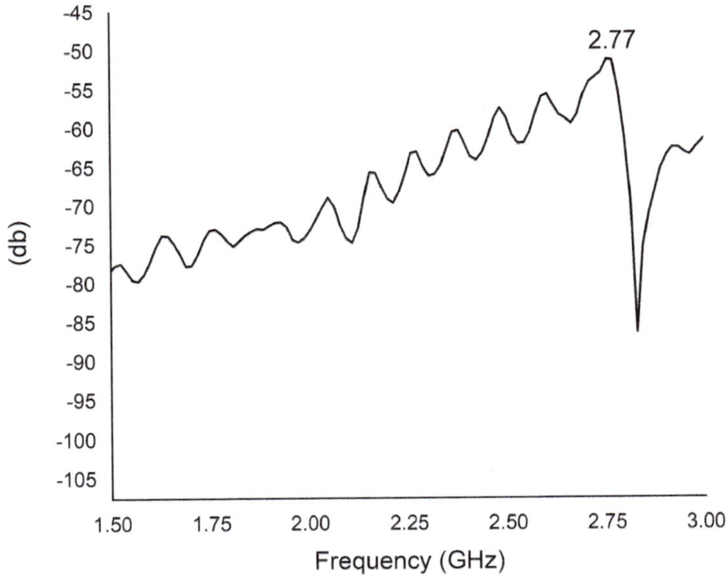

Fig. 12.12: Resonating frequency of ring resonator without exposure to stimuli.

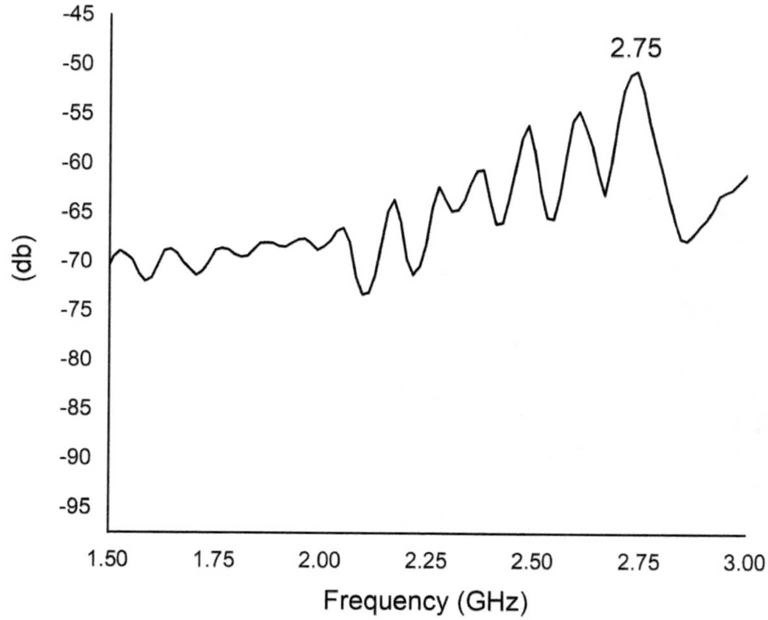

Fig. 12.13: Resonating frequency of ring resonator with exposure to stimuli (without moisture).

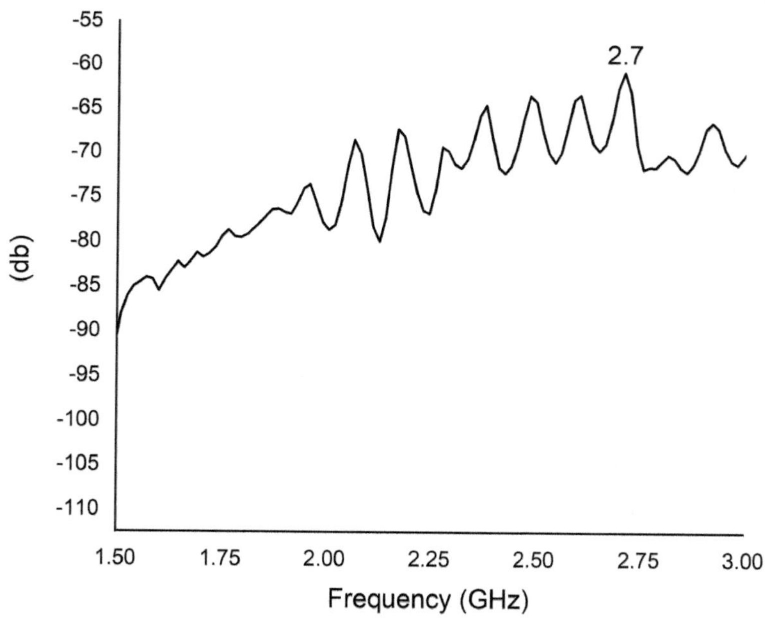

Fig. 12.14: Resonating frequency of ring resonator with exposure to stimuli (with moisture).

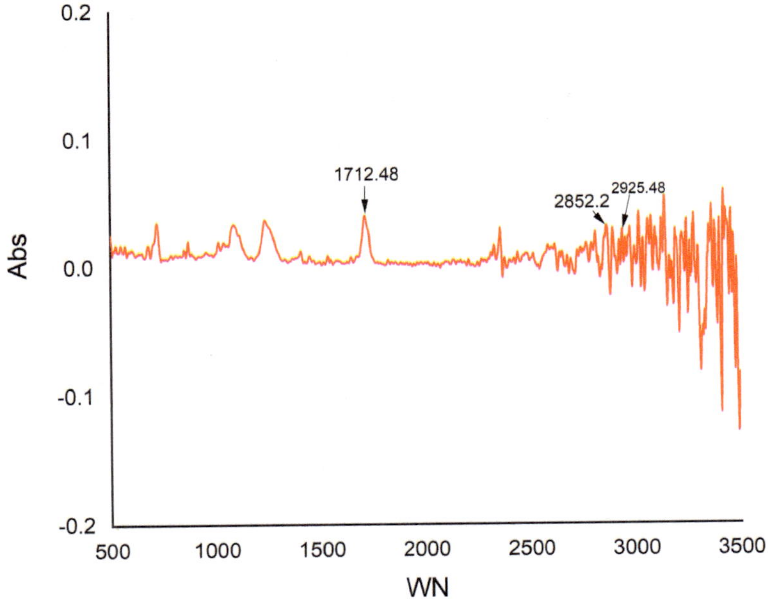

Fig. 12.15: FTIR of PETG without exposure to stimuli.

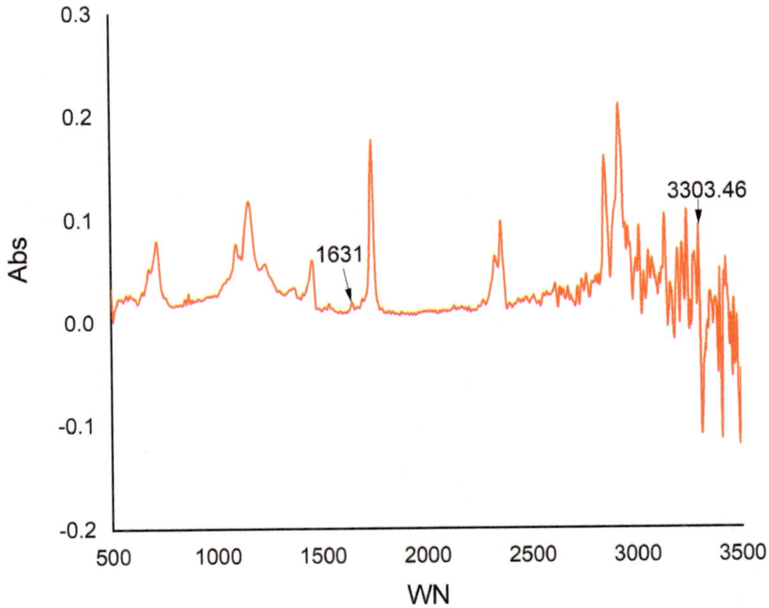

Fig. 12.16: FTIR of PETG with exposure to stimuli (without moisture).

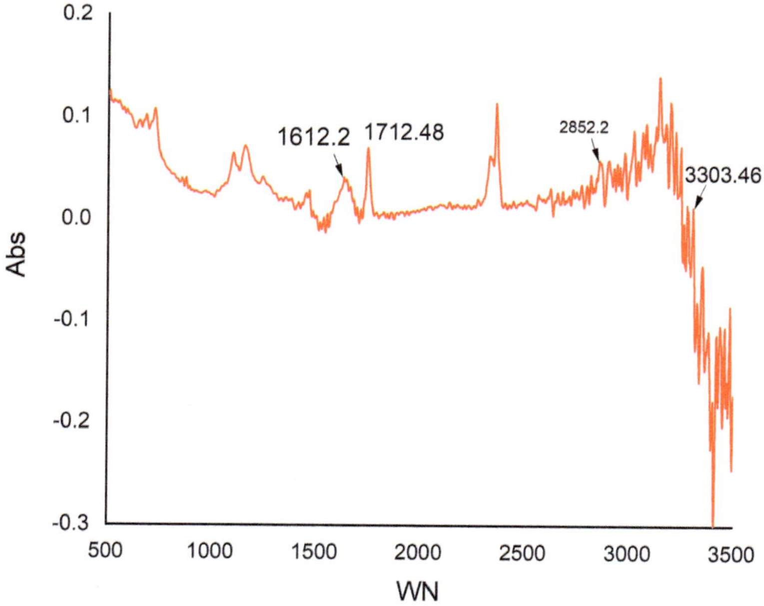

Fig. 12.17: FTIR of PETG with exposure to stimuli (with moisture).

12.12 V-R analysis

V-R characteristics of the substrate were obtained under different conditions, without exposure to stimuli, with exposure to stimuli (without moisture), and with exposure to stimuli (with moisture). These tests were performed on the source meter unit (Fig. 12.18).

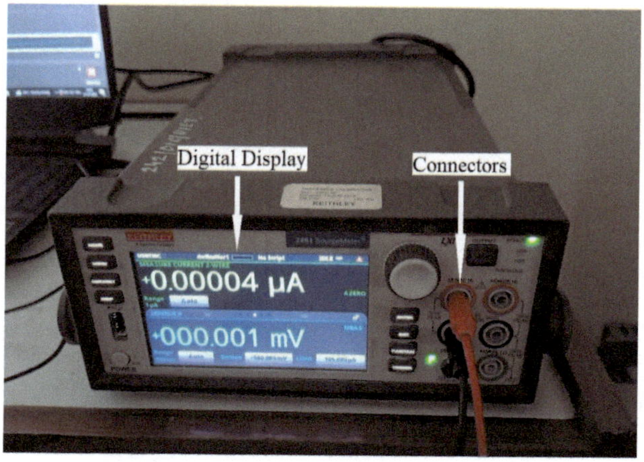

Fig. 12.18: Source meter unit.

The values of the resistance under 3 conditions are 58.24 Ω, 47.08 Ω, and 42.5 Ω for the PETG substrate without exposure to stimuli, with exposure of stimuli (without moisture), and with exposure to stimuli (with moisture) respectively (Fig. 12.19–12.21).

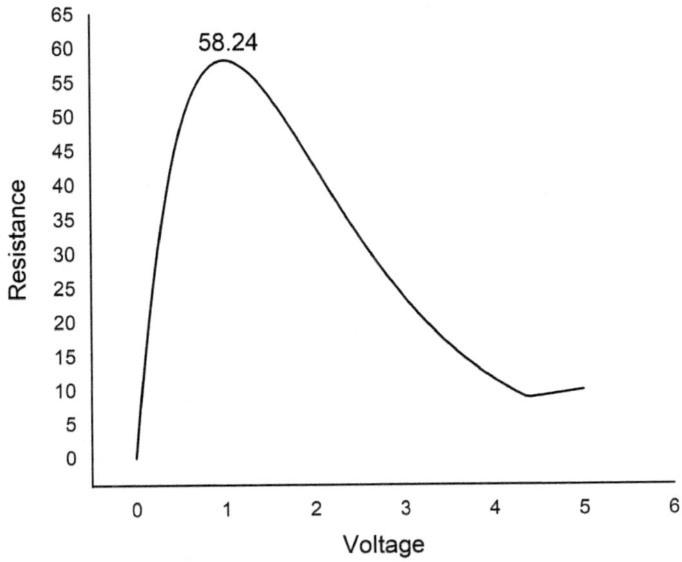

Fig. 12.19: V-R graph of PETG without exposure to stimuli.

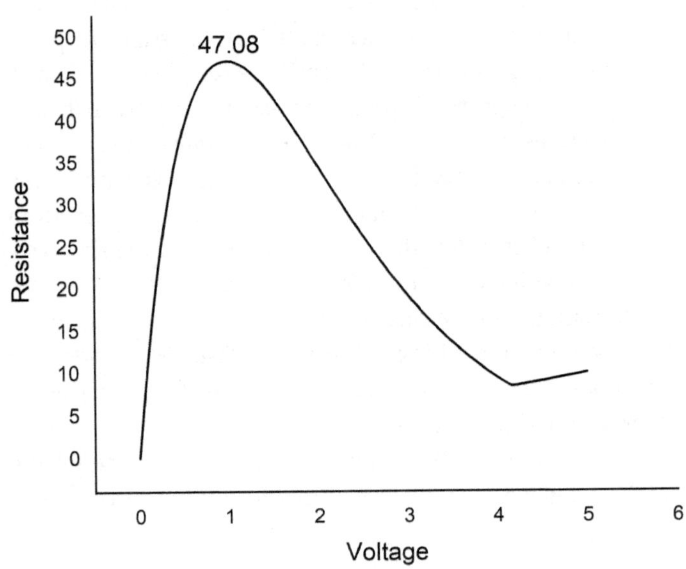

Fig. 12.20: V-R graph of PETG with exposure to stimuli (without moisture).

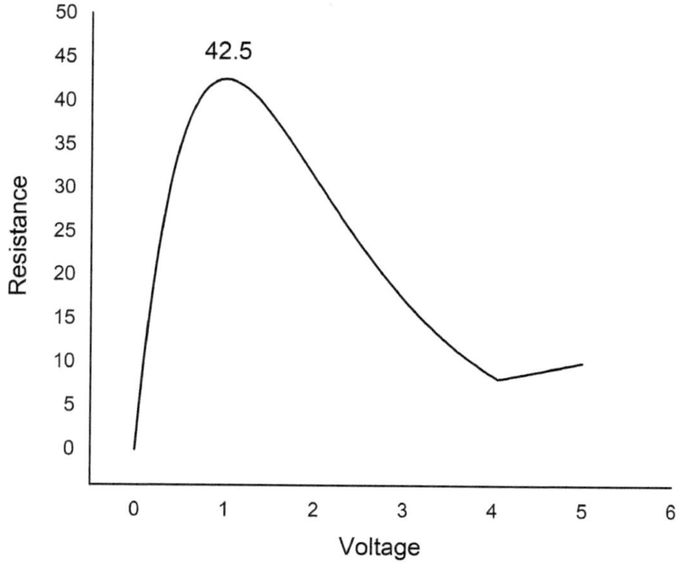

Fig. 12.21: V-R graph of PETG with exposure to stimuli (with moisture).

12.13 Conclusions

In this study, a substrate of waste PETG was 3D-printed, and a ring resonator was fabricated for the sensing applications in the smart container. In the FTIR spectrum, the peaks corresponding to the wavenumber 1,712.48 cm^{-1}, 2,852.2 cm^{-1}, 2,925.48 cm^{-1}, 1,631 cm^{-1}, 3,303.46 cm^{-1}, and 1,612.2 cm^{-1} with exposure of stimuli represents the PETG polymer, mustard oil, NaCl, and water molecules, respectively. The effect of stimuli on the RF characteristics was observed and it has been observed that the RF amplitude decreases as the ring resonator was exposed to the stimuli (with moisture). The RF decreased from 2.75 GHz to 2.7 GHz as the exposure of stimuli with moisture was given to the ring resonator as compared to the stimuli without moisture. Whereas, the value of ε was increased from 2.68 to 2.74. The resistance of the ring resonator with the effect of stimuli without moisture and with moisture decreased from 47.08 Ω to 42.5 Ω. The considerable changes in the dielectric properties of the PETG substrate may be tuned for sensing applications in the smart container. The change in the resistance of the PETG substrate can be used as an IoT-based solution when tuned with the Bluetooth of a smartphone for the online health monitoring of FA stored in smart containers. This also helps in ascertaining the shelf life of FAs.

References

Bałdowska-Witos P, Kruszelnicka W, Tomporowski A Life cycle assessment of beverage bottles. Journal of Physics: Conference Series 2020;1426(1):012038. IOP Publishing.

Balint T, Živčák J, Hudák R, Tóth T, Kohan M, Lancoš S Destructive and non-destructive testing of samples from PLA and PETG materials. IOP Conference Series: Materials Science and Engineering 2021 Nov 1;1199(1):012045. IOP Publishing.

Credi C, Griffini G, Levi M, Turri S. Biotinylated photopolymers for 3D-printed unibody lab-on-a-chip optical platforms. Small. 2018 Jan;14(1):1702831.

Epps III TH, Korley LT, Yan T, Beers KL, Burt TM. Sustainability of synthetic plastics: Considerations in materials life-cycle management. JACS Au. 2021 Dec 22;2(1):3–11.

Fan D, Li Y, Wang X, Zhu T, Wang Q, Cai H, Li W, Tian Y, Liu Z. Progressive 3D printing technology and its application in medical materials. Frontiers in Pharmacology. 2020 Mar 20;11:122.

Fina F, Goyanes A, Gaisford S, Basit AW. Selective laser sintering (SLS) 3D printing of medicines. International Journal of Pharmaceutics. 2017 Aug 30;529(1–2):285–93.

Guessasma S, Belhabib S, Nouri H. Printability and tensile performance of 3D printed polyethylene terephthalate glycol using fused deposition modeling. Polymers. 2019 Jul 22;11(7):1220.

Habel C, Schöttle M, Daab M, Eichstaedt NJ, Wagner D, Bakhshi H, Agarwal S, Horn MA, Breu J. High-barrier, biodegradable food packaging. Macromolecular Materials and Engineering. 2018 Oct;303 (10):1800333.

Harle L, Papapolymerou J. 3D printed membrane microstrip patch antenna. In2017 IEEE International Symposium on Antennas and Propagation & USNC/URSI National Radio Science Meeting. 2017 Jul 9, 2637–38. IEEE.

Huang L, An S, Li C, Huang C, Wang S, Zhang X, Xu M, Chen J, Zhou L. Performance of waste-paper/PETG wood–plastic composites. AIP Advances. 2018 May 1;8(5).

Kuswandi B, Wicaksono Y, Jayus, Abdullah A, Heng LY, Ahmad M. Smart packaging: Sensors for monitoring of food quality and safety. Sensing and Instrumentation for Food Quality and Safety. 2011 Dec;5:137–46.

Latko-Durałek P, Dydek K, Boczkowska A. Thermal, rheological and mechanical properties of PETG/rPETG blends. Journal of Polymers and the Environment. 2019 Nov;27:2600–06.

Lukianchuk I, Tulashvili Y, Podolyak V, Horbariuk R, Kovalchuk V, Bazyl SM. Didactic Principles of Education Students 3D-printing. 2022.

Mikula K, Skrzypczak D, Izydorczyk G, Warchoł J, Moustakas K, Chojnacka K, Witek-Krowiak A. 3D printing filament as a second life of waste plastics – A review. Environmental Science and Pollution Research. 2021 Mar;28:12321–33.

Moreno Nieto D, Alonso-García M, Pardo-Vicente MA, Rodríguez-Parada L. Product design by additive manufacturing for water environments: Study of degradation and absorption behavior of pla and petg. Polymers. 2021 Mar 26;13(7):1036.

Mukhopadhyay A. Process for delamination of laminated packaging. United States patent US 8,945,396. 2015 Feb 3.

Paszkiewicz S, Szymczyk A, Pawlikowska D, Irska I, Taraghi I, Pilawka R, Gu J, Li X, Tu Y, Piesowicz E. Synthesis, and characterization of poly (ethylene terephthalate-co-1, 4-cyclohexanedimethylene terephtlatate)-block-poly (tetramethylene oxide) copolymers. RSC Advances. 2017;7(66):41745–54.

Petrov P, Agzamova D, Pustovalov V, Zhikhareva E, Saprykin B, Chmutin I, Shmakova N. Research into the effect of the 3D-printing mode on changing the properties of PETG transparent plastic. 2021.

Reinaldo JD, Nascimento MC, Ito EN, Hage Junior E. Rheological, mechanical and morphological properties of poly (methyl methacrylate)/poly (ethylene terephthalate) blend with dual reactive interfacial compatibilization. Polímeros. 2015 Sep;25:451–60.

Revilla-León M, Özcan M. Additive manufacturing technologies used for processing polymers: Status and potential application in prosthetic dentistry. Journal of Prosthodontics. 2019 Feb;28(2):146–58.

Shamsuyeva M, Endres HJ. Plastics in the context of the circular economy and sustainable plastics recycling: Comprehensive review on research development, standardization, and market. Composites Part C. 2021 Oct 1;6:100168.

Singh K, Singh R, Singh AP. 3D-printed PETG-based smart containers for online health monitoring of food articles. National Academy Science Letters. 2023 Apr;46(2):113–16.

Singh R, Kumar S, Singh AP, Wei Y On comparison of recycled LDPE and LDPE–bakelite composite based 3D printed patch antenna. Proceedings of the Institution of Mechanical Engineers, Part L: Journal of Materials: Design and Applications. 2022 Apr;236(4):842–56.

Sölmann S, Rattenholl A, Blattner H, Ehrmann G, Gudermann F, Lütkemeyer D, Ehrmann A. Mammalian cell adhesion on different 3D printed polymers with varying sterilization methods and acidic treatment. AIMS Bioengineering. 2021;8(1).

Szykiedans K, Credo W, Osiński D. Selected mechanical properties of PETG 3-D prints. Procedia Engineering. 2017 Jan 1;177:455–61.

Tamoor M, Samak NA, Jia Y, Mushtaq MU, Sher H, Bibi M, Xing J. Potential use of microbial enzymes for the conversion of plastic waste into value-added products: A viable solution. Frontiers in Microbiology. 2021 Nov 30;12:777727.

Zhang S, Njoku CC, Whittow WG, Vardaxoglou JC. Novel 3D printed synthetic dielectric substrates. Microwave and Optical Technology Letters. 2015 Oct;57(10):2344–46.

Index

https://doi.org/10.1515/9783111081632-013